Consumer Reaction, Food Production and the Fukushima Disaster

Kentaka Aruga

Consumer Reaction, Food Production and the Fukushima Disaster

Assessing Reputation Damage Due to Potential Radiation Contamination

Springer

Kentaka Aruga
Graduate School of Humanities and Social Sciences
Saitama University
Saitama, Saitama
Japan

ISBN 978-3-319-59848-2 ISBN 978-3-319-59849-9 (eBook)
DOI 10.1007/978-3-319-59849-9

Library of Congress Control Number: 2017943183

Printed on acid-free paper

This Springer imprint is published by Springer Nature
The registered company is Springer International Publishing AG
The registered company address is: Gewerbestrasse 11, 6330 Cham, Switzerland

Preface

When the Tohoku-Pacific Ocean Earthquake occurred on March 11, 2011, I was working in an office located in Hayama, Kanagawa Prefecture. Although my office was more than 400 km apart from the epicenter of the earthquake, the quake was at a level that I have never experienced before. When I checked the news about the earthquake I learned that more than a 15 m high tsunami was approaching the coast of Tohoku regions but I first could not believe that this was really happening. I realized that the disaster was a reality after seeing entire communities swept away by these tsunamis and people desperately trying to evacuate from them on TV. However, towns and farmlands destroyed by the tsunamis were only the beginning of the disaster. The tsunamis had triggered another calamity. The No.1 reactor at the Fukushima Daiichi Nuclear Power Plant had exploded after hit by these massive waves. After this first accident was reported, the situation at the Fukushima plant grew gravely worse ultimately leading to a nuclear meltdown that emitted large quantities of radioactive material into atmosphere.

After the Fukushima disaster, the house and the town I lived at that time suffered from only a power outage, but this was nothing compared to the devastation that occurred in the Tohoku and other regions near the FDNPP. This destruction was beyond anyone's comprehension. In fact, even after six years of this catastrophe, there are still thousands from these regions who cannot return to their homes due to radioactive contamination.

The reason I started to work on the issue of reputation damage in regions suffering near the FDNPP is because I wanted to use my expertise to help these regions recover from this disaster. However, soon after I started the research, I realized the difficulty of the topic I was dealing with. Reputation damage is a problem that occurs because the individual's decision is often affected by the opinions of other individuals and it is easy to be influenced by false information. Furthermore, as the issue of radioactive contamination of food involves uncertainties, it was expected that there would be many people who will utterly avoid any products with bad reputation because they do not want to spend time searching about how safe the products are. Thus, whether the level is large or small, I noticed from the beginning that there would be reputation damage.

However, I did not want to connect all the avoiding behaviors of the consumers to reputation damage. This is why I tried to rely on the data to explain my views and tried not to argue the problem only from the perspective of reputation damage. This is the reason that the book resulted in having a mixed conclusion that the consumers' avoiding behavior toward the agricultural products of regions near the FDNPP is partly affected by false reputation and partly caused by factors not directly related to reputation. However, it was meaningful to reveal through the book what types of consumers are eager to buy products from these regions and to identify the factors affecting the consumers' reaction toward products from these regions.

In particular, it was interesting to find that consumers with high interest in environmental problems and helping the disaster-affected regions to restore from their damage have the tendency to buy products from regions near the FDNPP. It was apparent from this finding that altruistic consumers are more serious about helping the disaster-affected regions to recover their economy to the pre-crisis level.

If altruism is the important factor for the consumers to have a positive reaction toward products from regions near the FDNPP it might be that increasing the number of people that care about other people will be the solution for reputation damage. However, as investigating how altruism affects consumer behavior is beyond the scope of my specialty (economics), it might be useful to incorporate the methods used in psychology for further study.

The survey data used in this book was first intended to survey 6,000 people. Yet, when I announced to the registered members of the internet survey company about the survey, I was able to gather more than 8,000 respondents. Hence, I realized how interested people were about this issue.

Part of the outcomes of the book were presented at the 2015 conference of the International Association for Energy Economics held in Antalya, Turkey, 2015 conference of the Australian Agricultural and Resource Economics Society in Canberra, Australia, and 2016 conference of the European Association of Environmental and Natural Resource Economists in Zurich, Switzerland. Here too, I found that many people were interested in how consumers were reacting to the agricultural products from regions near the FDNPP after the Fukushima disaster.

Saitama, Japan Kentaka Aruga

Acknowledgements

I would like to note that this study is supported by JSPS KAKENHI Grant Number 25712025 titled, "Studying the Consumer Reaction toward Radioactively Contaminated Foods after the Fukushima Accident."

I would like to thank Suguru Isoda, Hiroki Ichihashi, and Ryusuke Maeda, students of the Ishikawa Prefectural University when I was writing the book, for helping me create the figures and tables used in this book. I am also grateful to Kumiko Matsui, the editor of Showado Inc., and Fritz Schmuhl, the editor of Springer for giving me the opportunity to write a book on this topic. Furthermore, I would like to express my gratitude to Glen Norris, an associate professor at Ishikawa Prefectural University for helping me edit the English of this book.

Contents

Chapter 1
Introduction

On March 4, 2011, a massive Tsunami surged along the Pacific coast of Tohoku region destroying the turbine building of the Fukushima Daiichi nuclear power plant (FDNPP). This accident led to a significant nuclear disaster that Japan had never experienced before. Soon after this catastrophe occurred, a substantial amount of radioactive material was released into the air, and various food products of regions near the FDNPP became contaminated with radiation. Such food includes vegetable, fruit, beef, fish, tea-leaf, mushrooms, and so on. The Japanese government immediately established an exacting safety standard to regulate the limits of radioactivity in food products as to prevent consumers from purchasing food contaminated with radiation.

Even after this safety standard in food commodities were enforced on all food products in Japan, there remained some problems. One is that it was difficult for the consumers to trust the national safety standard. The other problem is that even if the safety standard was trustable, there were possibilities that some dishonest distributors might sell food products that are contaminated with false reports and do not meet the safety standard. Thus, even after more than 5 years from the Fukushima disaster, there are still consumers who avoid buying food produced near the FDNPP.

In fact, one of the largest grocery retailers in Japan has found some radioactive materials in some food sold in its store after conducting a proper investigation of the level of radiation for its groceries and the store had to restrict selling this food. So, it is true that it is a tough task to eliminate the risk of consumers to have radiation-contaminated food in their hands. However, we also need to note that the producers of Fukushima are facing the so-called "reputation damage" problem: some food that has no risk of radiation contamination are not sold just because some media and false status have influenced the consumers to believe that anything produced near Fukushima is unsafe.

This book will identify the situation of the reputation damage after the Fukushima disaster using a consumer survey data which contains about 8700 respondents throughout Japan. The study was conducted for nine food products:

K. Aruga, *Consumer Reaction, Food Production and the Fukushima Disaster*,
DOI 10.1007/978-3-319-59849-9_1

rice, cucumbers, apples, *shiitake* mushrooms, beef, pork, eggs, tuna fish, and *wakame* seaweed. These products have a relatively high production volume in Fukushima prefecture. We also conducted a survey for mineral water because drinking water of some regions near the FDNPP became contaminated with radiation after the Fukushima disaster. The survey not only used respondents in Tohoku region and Tokyo metropolitan area who had actual damage from the Fukushima disaster but also those who live apart from the FDNPP and had no direct impact from the catastrophe.

There is similar consumer survey performed by the Consumer Affairs Agency, Government of Japan, which is also a survey designed to configure the condition of the reputation damage in food products produced near the FDNPP. However, this study does not conduct a numerical analysis on the survey data, and it only shows the aggregated results of the survey. It does not explain how and which type of consumer reactions are influencing the reputation of food produced near the FDNPP. There are still only a few studies that analyze the relationships between consumer response toward food of regions near the nuclear FDNPP and reputation damage using an extensive consumer survey data. Hence, more studies need to be conducted to understand the situation of reputation damage after the nuclear disaster.

Recently, there are some studies carried out by Japanese scholars that assess the status of the reputation damage after the Fukushima disaster. However, these studies only use survey data with respondents of Tohoku region and Tokyo metropolitan area who had direct effect from the Fukushima disaster. As far as we know, a survey data that includes respondents of all parts of Japan is still rare. Due to the recent development of the distribution technology such as the cold chain,[1] more producers are now selling their food products to all parts of Japan, and the Japanese food market is becoming integrated. Thus, when investigating the consumer reaction toward food produced near the FDNPP, it is becoming more important to conduct a survey to include consumers of all parts of Japan.

This book uses the contingent valuation method (CVM),[2] which is a standard tool to estimate the value of a nonmarket good. This method is often used in the field of environmental economics. This book uses this CVM to identify the consumer's willingness to buy a product in a virtual market that has the risk of nuclear contamination. The book will also configure the factors that affect the consumer's purchasing behavior in this hypothetical market. Finally, the book will investigate and clarify whether false reputation influences consumer reactions toward food produced near the FDNPP. This book will analyze how differences in consumer eating habits, perceptions on food safety, interests in social problems, attitudes

---

[1]Cold chain is a food supply chain system, where the temperature of a food product during its delivery is controlled at a constant level using the latest freezing and storing technology.

[2]CVM is a tool to estimate the value of the goods that do not have actual markets such as public parks, ecosystem services, and benefits of beautiful landscape. A survey is conducted to ask the respondents their willingness to pay for these goods in a virtual market. Then, their economic values are estimated using an economic model and econometric tools.

toward nuclear contamination, and their social attributes will affect their reaction toward food produced near the FDNPP. Then we will try to configure how their response is linked to the reputation damage.

Studies to understand the situation of the reputation damage through unfounded rumors for food produced in regions near the FDNPP will be very helpful for the people in these areas providing agricultural products to recover its sales to the pre-disaster level. The reputation damage that is occurring after the Fukushima disaster is closely related to consumer reaction toward information on the risk of radiation contamination in food. Hence, I believe this book provides a helpful resource for deciding which type of consumers will be a good target for increasing the sales and for constructing efficient marketing policy for such regions to recover their economies.

A nuclear disaster that led to a radiation contamination in the surrounding area like the Fukushima disaster is a rare case in the world, so there are still a few studies that investigate how consumers react toward food with the potential risk of radiation contamination and what factors affect consumer's decision when purchasing such food. Thus, this book is a valuable case study for understanding the consumer behaviors in a situation where food becomes contaminated with radiation after a nuclear disaster.

There are not many books that deal with the food safety problem from a consumer viewpoint that utilizes consumer survey data. Therefore, I recommend this book to students and consumers who are concerned with food safety issues, policy makers who are dealing with reputation damage in the agricultural industry, and people in the food industry that deal with consumer relations.

Chapter 2
Radiation Contamination of Agricultural Products

The Fukushima nuclear disaster started when the Tohoku-Pacific Ocean Earthquake hit the Tohoku and Kanto region at *14:46* JST on *March 11*, 2011. The epicenter of this earthquake was located about 130 km off the coast of Oshika Peninsula of Miyagi Prefecture, and was about 24 km below sea level. The scale of the quake was estimated to be magnitude 9.0, which was the greatest earthquake ever recorded in Japan.[1] It is still vivid in our memory that the effect of the earthquake had devastating impacts on not only in the Tohoku region, but also in the Tokyo metropolitan area located more than 300 km apart from the epicenter. According to the Asahi newspaper, in the 12 prefectures, about 15,890 people died from the catastrophe, and in the six prefectures, 2589 people became missing as of February 2015.

The Tohoku-Pacific Ocean Earthquake, later named the Great East Japan Earthquake, or simply 3.11, after the disaster. The big difference of this earthquake from other large earthquakes that occurred in the world is that the earthquake led to a major nuclear accident. There are several earthquakes in the world where the earthquake triggered a tsunami like the 2004 Indian Ocean earthquake and the 2006 Pangandaran earthquake of West Java.[2] However, the Tohoku-Pacific Ocean Earthquake is the world's first earthquake where the tsunami triggered by an earthquake led to a nuclear accident.

---

[1]Magnitude scales used in seismology represents the size of an earthquake as the power of energy using a logarithmic function.

[2]The 2004 Indian Ocean earthquake occurred on the morning of December 26. Its epicenter was located off the west coast of Sumatra, Indonesia. After the earthquake, the coastlines of Indonesia, Thailand, and Sri Lanka were hit by a tsunami. About 230 thousand people died or became missing due to this tsunami. It is reported that the maximum height of the tsunami was greater than 30 m (NILIM 2005).

The 2006 Pangandaran earthquake occurred in the evening of July 17. The earthquake had a moment magnitude of 7.7 and the epicenter was along the southern coast of the Island of Java, Indonesia. It is known that more than 600 people died from a tsunami that was triggered by this earthquake. The maximum height of the tsunami is estimated to be over 5 m (Nanayama et al. 2007).

K. Aruga, *Consumer Reaction, Food Production and the Fukushima Disaster*,
DOI 10.1007/978-3-319-59849-9_2

This chapter will provide a brief overview of the Fukushima Daiichi nuclear power plant (FDNPP) accident and the situation of the radiation contamination of agricultural products after the nuclear disaster.

2.1 The Fukushima Daiichi Nuclear Power Plant Accident

2.1.1 Overview of the Accident

The FDNPP accident started when a magnitude 9.0 earthquake, whose epicenter was offshore Sanriku occurred at 2:46 pm on March 11, 2011. After this first earthquake, more than eight earthquakes with a seismic intensity of 5 or greater occurred near the FDNPP. At the FDNPP, the transmission steel towers collapsed and power lines were cut and the plant lost its all seven electricity transmission lines. Moreover, the FDNPP lost its connection with the external power source, which led to a station blackout.

Every nuclear power plant is installed with an emergency diesel generator (EDG). When the power plant loses its external power source, the power will be supplied from this EDG. Right after the Tohoku-Pacific Ocean earthquake occurred, the EDG was running normally at the FDNPP and supplying electricity to the plant. However, after 45 min of the first earthquake, the first wave of tsunami hit the nuclear plant and the second wave of tsunami that was about 13 m high continuously hit the nuclear plant. This tsunami watered the four reactor units of the FDNPP, and this turned off the function of the EDG. This is how the FDNPP lost all the power supply for its four reactor units, and this blackout situation continued for 10 days.

As Unit 1 through Unit 4 reactors of the FDNPP lost their power, the cooling function of the reactor stopped working and this triggered the reactor core to start melting. And finally, due to this nuclear meltdown, radioactive materials that were trapped in the fuel pellet were released into the air along with the water vapor. This is how the FDNPP accident led to a nuclear disaster causing a radioactive contamination of its surrounding areas.

Thus, although the main cause of the FDNPP accident is the earthquake and the tsunami, the reason for this accident to become a nuclear disaster is attributed to the station blackout that lasted for ten days. It is suggested that if the electricity recovered earlier, the plant could have controlled the reactor from melting (Ishikawa 2014).

On the day of the earthquake, the Fukushima Daini nuclear power plant, which was only about 12 km apart from the FDNPP (see Fig. 2.1), was also hit by the tsunami.[3] The Fukushima Daini power plant was also drown with water after the tsunami arrived, but it did not become a disaster that led to a nuclear meltdown.

[3]"Daiichi" and "Daini" means the No.1 and No. 2 in Japanese and as seen in Fig. 2.1 the Fukushima Daiichi and Daini nuclear plants were located very close (about 12 km) to each other.

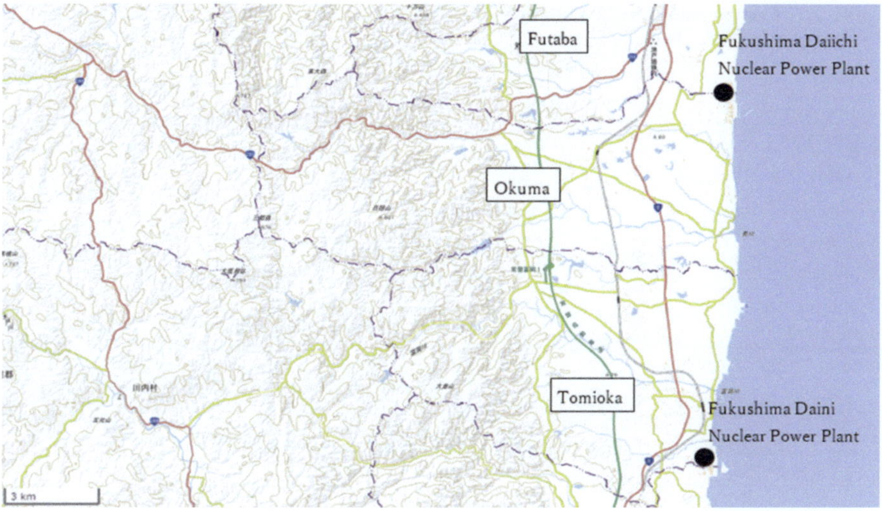

Fig. 2.1 Location of the Fukushima Daiichi Nuclear Power Plant and the Fukushima Daini Nuclear Power Plant (Reproduced from Aruga 2016)

According to Ishikawa (2014), it is believed that the Fukushima Daini nuclear power plant had electricity even after the tsunami because the EDG continued to work normally and was able to cool down the nuclear reactor. Hence, the reason why the FDNPP led to a nuclear meltdown is due to losing all its electricity supply.

2.1.2 Degree of the Nuclear Accident

Here, we would like to compare the extent of the Fukushima nuclear accident with the 1986 Chernobyl nuclear accident so that we can understand the severeness of the nuclear accident compared with the previous accidents of the world. The Nuclear Safety Commission (NSC) [the current Nuclear Regulation Authority (NRA)] of Japan has estimated the total amount of radioactivity released from the Fukushima accident to be 0.77 quintillion Bq. As seen in Table 2.1, the estimated amount of radioactivity released from the Fukushima accident is around one-seventh of the amount released from the Chernobyl disaster (NERH 2011). However, the Fukushima accident has been rated level 7 on the International Nuclear and Radiological Event Scale (INES), which is the highest level within the INES scale and the same scale as the Chernobyl accident.[4] Hence, it is certain that the FDNPP accident is among one of the worst nuclear disasters in the world.

[4]INES is an international index used to describe the relative magnitude of a nuclear accident. The scale is established by the IAEA (International Atomic Energy Agency) and the OECD.

Table 2.1 Comparison of the amount of radioactivity released from the Fukushima and Chernobyl accidents (*Unit* 10^{16} Bq)

	Physical half-life time	Fukushima accident	Chernobyl accident
Iodine-131	8 days	16	180
Cesium-134	2 years	1.8	4.4
Cesium-137	30 years	1.5	8.5
Strontium-90	29 years	0.014	0.8
Plutonium-239	24 thousand years	0.003×10^{-4}	0.003
Total emission in Iodine equivalent	–	77	520

Source Aruga (2016, p. 4)

2.1.3 Meaning of the Level 7 INES Scale

How serious is the level 7 INES scale that was rated for the Fukushima nuclear disaster? Figure 2.2 illustrates the naming of the INES level 1 through 7 scales. As seen in the figure, the INES use the word "incident" for the levels 1–3 nuclear events while the word "accident" is used for the levels 4–7 events. One level of

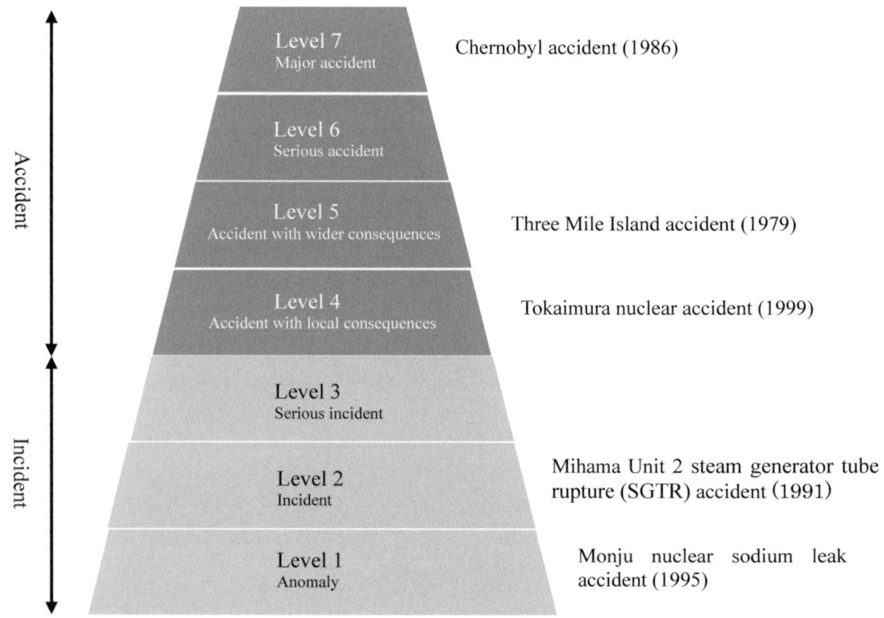

Fig. 2.2 International Nuclear and Radiological Event Scale (INES). *Source* Aruga (2016, p. 5)

increase in this scale represents that the nuclear event is about ten times more severe than the previous level. The difference between the INES incident and accident is whether there was at least one death from radiation by the nuclear event. The levels 1–3 nuclear events use the word “incident” because there was no death from the incident and the event only caused an injury to some person. On the other hand, in the levels 4–7 scales that are categorized by the word “accident,” at least one person has died from the nuclear event.

The Fukushima nuclear disaster was rated the level 7 scale, which means that the disaster was rated the highest level of scale among the 7 INES scales. The criteria for the level 7 scale is based on whether the event is “resulting in an environmental release corresponding to a quantity of radioactivity radiologically equivalent to a release to the atmosphere of more than several tens of thousands of terabecquerels of iodine-131” (IAEA 2008 p. 28).

The Fukushima disaster did release an enormous amount of radioactive materials to its surrounding areas so that the people who lived within the 20 km radius from the FDNPP were forced to evacuate after the nuclear accident and some of these people were never able to return to their houses. In this sense, it is certain that the Fukushima disaster was one of the worst nuclear accident ever happened in the world.

2.1.4 The Number of Evacuees After the Accident

After April 2011, regions within the 20 km radius from the FDNPP was designated as the warning zone and entering this zone was restricted and prohibited. These zones have been established to prevent the people living near the FDNPP from getting exposed to high level of radiation. After this zone was built, about 78,000 people had to evacuate from the area.

Besides, regions that are 20–30 km radius distance from the FDNPP was selected as the emergency evacuation preparation zone. People who belonged to this zone had to either evacuate or prepare for evacuation. There were about 58,510 people who evacuated from this zone.

Finally, for regions more than 20 km radius distance from the FDNPP that had a high risk of getting exposed to radiation was chosen as the planned evacuation zone. The high risk here meant the expected amount of radiation dose in the area was more than or equal to 20 mSv per year. People in this zone had to evacuate immediately and the number of evacuees in this zone was about 10,010.

By adding all the evacuees from these zones, there were a total of about 146,520 evacuees (see Table 1.2) who had to evacuate from their houses after the accident (Table 2.2).

Table 2.2 Evacuation zones and the number of evacuees per zone

Zone	Distance from the FDNPP	Number of evacuees
Warning zone	Within 20 km radius distance	About 78,000
Emergency evacuation preparation zone	20–30 km radius distance	About 58,510
Planned evacuation zone	More than 20 km radius distance but high risk of radiation exposure	About 10,010
Total		About 146,520

Source Aruga (2016, p. 7)

2.2 Basics of Radiation

In this section, we first explain some basic terms that are relevant to radiation and are important when understanding the effects of radiation on food products. Then, we discuss the food safety standards of protection against radiation in Japan. Finally, we investigate the differences of the levels of radioactive contamination among the different types of agricultural products by showing which products contained radioactive materials above the national food safety standard after the Fukushima disaster.

2.2.1 Radiation, Radioactive Materials, and Radioactivity

A radiation in general, is a particle or wave that transmits or emits high energy through space at a high velocity. A radioactive material is a material that releases radiation. Moreover, radioactivity is the ability and characteristic of an unstable atomic nucleus to transform into stable or unstable products while emitting radiation.

Figure 2.3 illustrates these three words by metaphorizing them to a firefly, a firefly's light, and the ability of the firefly to release light from its body. As seen in the figure, if the firefly represents radioactive material, radiation will be the firefly's light, and radioactivity will be the ability of the firefly to release light from its body. The difference between a firefly and a radioactive material is that while the activities of a firefly have no negative influence to us, those of a radioactive material can become harmful and is invisible. Although there is a way to visualize a radioactive material using a special camera, a radioactive material is a very tiny particle and it is normally impossible to see with our eyes even if it was on the ground or was attached to some buildings or animals. Thus, there is a danger of being exposed to radiation without noticing it.

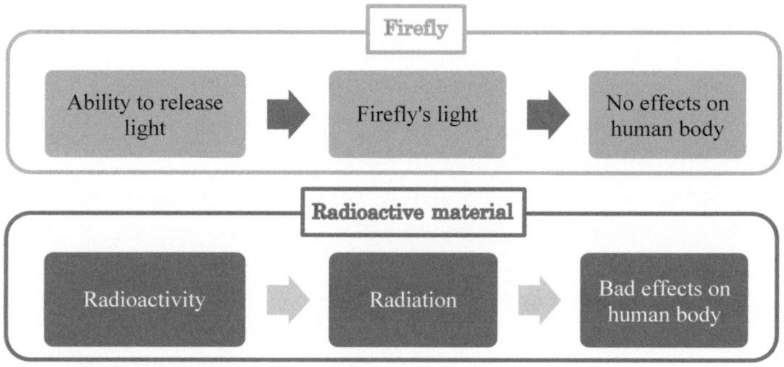

Fig. 2.3 Relations between radiation and radioactivity

2.2.2 *Radiation Exposure*

Radiation exposure means that the human body is exposed to radiation, but we should note that radiation exposure involves internal and external exposures. The external radiation exposure occurs when our body is exposed to a penetrating radiation field from outside our body. On the other hand, internal radiation exposure arises when we eat or drink food that contains a radioactive material and this material releases radiation inside the body.

As the word radiation exposure reminds us of how people suffered after the atomic bombing of Hiroshima and Nagasaki, many of us may think radiation exposure is by no means critical to the human body. However, we should know that we have always been exposed to radiation from nature since long time ago. It just did not harm us because the amount of the radiation dose we get from nature is usually subtle. Such radiation that exists in nature is called the natural radiation, and we are all exposed to a certain level of this natural radiation every year.

Then why are we so cautious about radiation exposure? That is because when the amount of radiation dose we receive is greatly increased it can have adverse effects on our body. Hence, to understand what amount of radiation dose can be harmful to our body, we will next look into the units of measurement used to configure the radiation dose.

2.2.3 *Units of Radiation Dose*

First, the unit becquerel (Bq) is a unit to measure the degree of radioactivity. A radioactive material will spontaneously emit energy as a result of the radioactive decay and becquerel measures the level of this emitted energy. To explain the meaning of this unit in terms of the firefly light, becquerel would be the level of the intensity of a light released from the firefly.

Becquerel is named after the French physicist Henri Becquerel, who discovered the radioactivity in 1896 and shared a Nobel Prize with Pierre Curie and Marie Curie. Becquerel indicates the number of radioactive nuclei that disintegrates per second in a radioactive material. For example, if 400 nuclei decay in 10 s, the level of the radioactivity for this material will be 40 becquerels. Becquerel is often used to show the degree of radioactivity per weight or volume such as becquerels per kilogram (Bq/kg), becquerels per liter (Bq/l), and becquerels per cubic meter (Bq/m^3).

Another important unit used to describe the intensity of radiation is the unit sievert. Sievert (Sv) is a unit to measure the health effect of ionizing radiation on a human body. Simply, it represents how much a human body can be affected when exposed to radiation. Sievert is named after a Swedish medical physicist Rolf Maximilian Sievert, who is renowned for his work on radiation dosage measurement and the biological effects of radiation during the nineteenth century. As sievert is a enormous unit, sievert is often represented in millisievert (mSv), one thousandth of sievert, or microsievert (μSv), one millionth of sievert. Sievert is often used to show how many levels of radiation one has received in a certain period such as mSv/year. For example, the effect of a radiation dose from a natural radiation on an average person is 2.4 mSv per year (RSC 2013) and this can be expressed as 2.4 mSv/year.

2.2.4 *Radiation Dose Limit*

The radiation does limit in Japan is now set to 1 mSv per year. This limit does not include the radiation dose from nature in a year. Although there is an argument whether or not this 1 mSv per year radiation dose is scientifically an adequate amount, on Nov. 11, 2011, the Japanese government established a regulation to set the annual limit of the radiation dose a person can receive besides the natural radiation to be 1 mSv. This radiation dose limit was the advisory level established by the International Commission on Radiological Protection (ICRP) (ICRP 2007).

2.2.5 *Japanese Food Safety Standards of Radioactive Cesium*

Based on the 1 mSv upper limit of radiation dose, the Japanese government has set safety standards for radioactive cesium in food as shown in Table 2.3. The unit of the cesium limit in the table is 1 becquerel of cesium per 1 kg of food. The reason for using cesium as the radiation safety standards is that as seen in Table 1.1, the physical half-life of cesium is relatively long and it is known that a significant amount of cesium-137 was released into air after the Fukushima disaster.

Table 2.3 Japanese food safety standards for radioactive cesium (Bq/kg)

Until March 2012 (Radiation dose limit: 5 mSv)		After April 2012 (Radiation dose limit: 1 mSv)	
Category	Limit	Category	Limit
Drinking water	200	Drinking water	10
Milk and dairy products	200	Raw milk	50
Vegetables	500	Infant food	50
Grains	500	General food products	100
Meat, eggs, fish, and others	500		

Source Aruga (2016, p. 11)

Until March 2012, the safety standards for radioactive cesium in food was limited to the total of 5 mSv per year in Japan but to enforce the safety standards the Japanese government established a more stringent standard in April 2012 and set the radiation dose limit to 1 mSv per year.

The numbers in Table 2.3 provided for different types of food represents the upper limit of radioactivity for these foods. The limit means that if the radioactivity in the food is below this level the effect of the radioactivity on the human body will stay within the 1 mSv per year level even if one continues to eat this food on a daily basis.

2.2.6 *International Comparison of the Safety Standards for Radioactive Cesium in Food*

Comparing the current food safety standards for radioactive cesium in Japan with other parts of the world, you can see from Table 2.4 that Japan has a comparatively strict standard. The reason for the European Union (EU) and the United States (US) having a lower limit of Bq/kg compared with Japan is because these EU and US standards are set under a normal condition having very low risk of distributing food contaminated with radioactive materials. Japan had to set a severe standard compared with these regions because the risk of distributing food with radioactive materials became very high after the Fukushima disaster.

The EU and the US soon restricted importing food products from Japan after the Fukushima disaster if the products did not meet the Japanese safety standards set

Table 2.4 Food safety standards for radioactive cesium in Japan, the EU, and the US(Bq/kg)

	Japan (current)	EU	US
Drinking water	10	1000	1200
Milk and dairy products	50	1000	1200
Infant food	50	400	1200
General food products	100	1250	1200

Source MHLW (2012)

after April 2012. Moreover, they prohibited to import any food products that are produced in regions near the FDNPP. Thus, although the food safety standards for the EU and US had lower standards than the Japanese standards, these standards only applied to their domestic products, and it is not possible to say that the Japanese standards are more severe compared with those of other countries.

In the next session, I would like to show how much radioactive materials were detected in different types of agricultural products and compare the level of radioactivity for these products with the Japanese safety standards as indicated in Table 2.3.

2.3 Conditions of the Radioactive Contamination of Agricultural Products After the Fukushima Disaster

After the Fukushima disaster, nuclear materials have been detected in various agricultural products such as rice, vegetables, fruit, mushrooms, and edible wild plants. These agricultural products became contaminated with radioactivity because radioactive materials adhered to plant leaves and soils, and these plants absorbed radioactive substances from their leaves and stems. Thus, high level of radioactive materials has been identified in products such as spinach whose product form is leaf. Fruit products such as Japanese plum (ume), Japanese citrus (yuzu), and loquat where the fruiting began in March also became highly contaminated because of the Fukushima disaster occurred in March. The soils of areas near the FDNPP had a serious level of contamination because the radioactive cesium, having a long physical half-life, stays in the soil for a long time.

For forest products, as fungal species are known to accumulate radiocesium (Yamada 2013), a high level of radioactive materials has been found in wild mushrooms. In particular, when the hardwoods used for cultivating wood mushrooms are contaminated with radioactive materials, the mushrooms harvested from this hardwood have a high probability of being contaminated. Hence, shipments of wild mushrooms from Tohoku regions, Ibaraki prefecture, and Tochigi prefecture were prohibited by the government after the Fukushima disaster.

2.3.1 Conditions of the Radioactive Contamination of Rice, Vegetables, Fruit, and Forest Products

Here I would like to explain the amounts of radioactive cesium detected in rice, vegetables, fruit, and forest products such as mushrooms and wild edible plants by comparing these amounts with the Japanese food safety standards for radioactive cesium.

Table 2.5 Percentages of test samples above the safety standards for rice, vegetables, fruit, and forest products

	Rice			Vegetables		
	Number of test samples	Between 50 to 100 Bq/kg (%)	Above 100 Bq/kg (%)	Number of test samples	Between 50 to 100 Bq/kg (%)	Above 100 Bq/kg
2011	26,464	3.1	2.2	12,671	1.2	3.0
2012	About 10.37 MM	2.0×10^{-4}	8.1×10^{-6}	18,570	5.4×10^{-4}	2.7×10^{-4}
2013	About 11.04MM	7.4×10^{-5}	2.5×10^{-6}	19,657	1.0×10^{-4}	0
	Fruit			Mushrooms and edible wild plants		
	Number of test samples	Between 50 to 100 Bq/kg (%)	Above 100 Bq/kg (%)	Number of test samples	Between 50 to 100 Bq/kg (%)	Above 100 Bq/kg (%)
2011	2732	7.5	7.7	3856	7.7	20.2
2012	4478	1.3	0.3	6588	8.6	9.2
2013	4243	0.7	0	7581	4.2	2.6

Source Aruga (2016, p. 13)

Table 2.5 is created based on the radiation test conducted on the agricultural products of 17 prefectures that are designated by the Japanese government to perform radiation tests for their products. These 17 prefectures consist of Fukushima, Ibaraki, Tochigi, Gunma, Chiba, Kanagawa, Miyagi, Iwate, Aomori, Akita, Yamagata, Niigata, Nagano, Saitama, Tokyo, Yamanashi, and Shizuoka. The percentage numbers in the table represent the ratio of test samples that contained radiation above the Japanese food safety standard for radiation. As the agricultural products in the table all belong to the general food products category in Table 2.3, any test samples whose level of radioactive cesium was above the 100 Bq/kg safety standard were included in these percentages. Table 2.5 also shows the percentages of the test samples whose radioactive cesium dose were more than 50 Bq/kg and less than or equal to 100 Bq/kg.

For the radiation test of rice conducted in 2011, 592 out of the 26,404 test samples (2.2% of the whole test samples for rice) contained radioactive cesium above the 100 Bq/kg safety standard (see Table 2.5). The number of test samples above the safety standard for rice decreased dramatically in 2012 and 2013, but this figure did not become zero. Thus, there were samples whose level of radiation cesium was above the safety standard even after one or two years from the Fukushima disaster.

For vegetables, 3% of the whole test samples contained radioactive cesium above the 100 Bq/kg safety standard.[5] Among the highly consumed vegetables in

[5]The definition of vegetable here is based on the classification of the Ministry of Agriculture, Forestry, and Fisheries of Japan.

Japan, cabbage, tomatoes, lettuce, Japanese green onion, broccoli, spinach, strawberries, macrophyll, perillas, and Chinese chives were the ones whose level of radioactive cesium exceeded the safety standard. On the other hand, onions, Japanese white radish (daikon), carrots, and cucumbers were the ones whose radiation dose of the test samples were all below the 100 Bq/kg safety standard. As seen in Table 2.5, percentages of the test samples of vegetables whose radiation dose was above the safety standard became tiny in 2012, and in 2013, none of the test samples in the vegetables contained radiation dose above the safety standard.

Compared with rice and vegetables, fruit had a high percentage of test samples that contained radioactive cesium above the food safety standard in 2011. However, the high percentage only lasted for a year, and as seen in Table 2.5, the percentage of samples above the safety standard became 0.3% in 2012. In 2013, none of the test samples had radioactive cesium above the safety standard. The fruit that was found to be contaminated with radiation includes Japanese plums, persimmons, kiwifruit, blueberries, mandarin oranges, peaches, prunes, Japanese citrus (yuzu), loquats, chestnuts, and so on. It is suggested that these fruits became contaminated with radiation because these fruits normally start to bear fruit in March where the Fukushima disaster occurred. In fact, for fruits whose fruiting occurs after May like apples and pears did not contain radioactive materials above the safety standard in their test samples.

Finally, mushrooms and edible wild plants had the highest percentage of test samples that contain radioactive cesium above the safety standard. As seen in Table 2.5, a high level of radioactive cesium was detected in mushrooms and edible wild plants even after 2 years from the Fukushima disaster compared with other agricultural products in the table. It is believed that these products are especially prone to radioactive contamination because fungi like the shiitake mushrooms have the characteristics of absorbing radioactive cesium.

2.3.2 Conditions of the Radioactive Contamination for Livestock Products

In this session, I would like to explain the conditions of the radioactive contamination for livestock products. After the Fukushima disaster, radioactive cesium has been detected in various livestock products. It is suggested that this contamination is attributed to the use of fermented rice straw as animal feed because the straws used for the feed were from the rice harvested in 2011 near the FDNPP (Aruga 2016). The Japanese government has set a provisional tolerance standard for animal feed to restrict the use of animal feed that contain radioactive cesium above this standard level to cope with the problem (see Table 2.6).

Radioactive contamination of livestock animals also occurs from the drinking water, so the Japanese government has advised the farmers to implement prevention measures such as to put a lid on the water tank.

Table 2.6 Provisional standards of radioactive cesium for Animal feed

Type of animal	Provisional standards (Bq/kg)
Calf and horse	100
Pig	80
Poultry	160

Source CAA (2014)

Next, I will explain the conditions of the radioactive contamination for the livestock products by comparing the amount of radioactive cesium detected in these products with the Japanese food safety standards.

First, for beef and pork products, although the percentage of the contaminated samples were less than 1% of the whole test samples, some of the test samples for these products contained radioactive cesium in 2011 (see Table 2.7). Only few test samples were identified to be contaminated in 2012, and in 2013, none of the samples contained radioactive cesium above the safety standard.

Second, I would like to discuss the level of contamination of raw milk. The food safety standard for raw milk is set to 50 Bq/kg to consider the amount of milk intake for infants. As seen in the table, the percentage of test samples for raw milk that contained radioactive cesium above the safety standard was less than 1% of the whole test samples in 2011. After 2012, none of the test samples for raw milk exceeded the national safety standard.

Finally, for chicken eggs, none of the test samples were found to be above the food safety standard even in 2011, and the risk of containing radioactive cesium has been low among the livestock products.

All in all, the potential of radioactive contamination disappeared within 1 year from the Fukushima disaster for all livestock products discussed in this session, and

Table 2.7 Percentages of test samples above the safety standards for livestock products

	Beef			Pork		
	Number of test samples	Between 50 to 100 Bq/kg (%)	Above 100 Bq/kg (%)	Number of test samples	Between 50 to 100 Bq/kg (%)	Above 100 Bq/kg (%)
2011	92,683	N/A	1.2	529	N/A	1.1
2012	148,644	6.7×10^{-6}	4.0×10^{-5}	773	2.6×10^{-4}	1.3×10^{-4}
2013	163,767	2.4×10^{-5}	0	619	0	0

	Raw milk		Chicken eggs		
	Number of test samples	Above 50 Bq/kg (%)	Number of test samples	Between 50 to 100 Bq/kg (%)	Above 100 Bq/kg (%)
2011	1919	0.42	419	N/A	0
2012	2421	0	464	0	0
2013	2040	0	311	0	0

Source Aruga (2016, p. 16)

it is likely that their level of the contamination was lower than the agricultural commodities explained in the previous session.

2.3.3 Conditions of the Radioactive Contamination of Seafood Products

In this session, we will investigate the conditions of the radioactive contamination of seafood products.

After the FDNPP accident, a large quantity of radioactively contaminated water spilled into the ocean and a high level of radioactive cesium has been detected in some fish caught near the FDNPP. Table 1.8 illustrates the percentage of test samples for seafood products that contained radioactive cesium above the level of food safety standards. As seen in the table, the percentage of test samples for Fukushima prefecture that surpassed the food safety standard was extremely high in 2011. Although the percentage decreased in 2012, nearly 3% of the whole test samples for Fukushima Prefecture detected a high level of radioactive cesium even in 2013. The Japanese government prohibited distribution of 28 seafood products that are caught near the coast of FDNPP, which include greenling, stone flounder, Japanese black porgy, Japanese halibut, and sea bass to cope with this situation.

Test samples conducted for seafood products outside the Fukushima Prefecture had a much lower percentage of test samples that contained radioactive cesium above the food safety standards (see Table 2.8). As seen in Table 2.8, the percentage in 2013 for seafood products outside the Fukushima prefecture was less than 1%.

Table 2.8 Percentages of test samples above the safety standards for seafood products

	All seafood products of Fukushima prefecture				All seafood products outside the Fukushima prefecture			
	Number of test samples		Above 100 Bq/kg (%)		Number of test samples		Above 100 Bq/kg (%)	
2011	3606		34.7		4970		4.5	
2012	6917		12.7		12,648		1.7	
2013	8521		2.8		12,174		0.5	
	Salmon		Bluefin tuna		Flounder		Wakame seaweed	
	Number of test samples	Above safety standards	Number of test samples	Above safety standards	Number of test samples	Above safety standards	Number of test samples	Above safety standards
Mar 2011–Aug 2015	438	0%	37	0%	4139	2.2%	571	0%

Source Aruga (2016, p. 17)

Finally, I would like to discuss the conditions of radioactive contamination among different types of seafood products. The results of the tests conducted for these samples are gathered from all parts of Japan and are freely available to the public. As seen in the table, highly migratory fish species like salmon and bluefin tuna did not have any test samples that exceeded the safety standard. Wakame seaweed, which is one of the most popular seafood products of the Tohoku region also did not contain test samples that were above the safety standard since the test began in March 2011. However, in flounder, which is a fish that live near the sea bottom, 2.2% of the test samples held radioactive cesium above the safety standard.

It is suggested that the reason for the fish that live at the sea bottom to have a high level of radioactive cesium is because radioactive materials that were spilled into the ocean accumulates at the bottom of the ocean. However, it is also known that radioactive cesium tends to be absorbed by clay minerals and that it is less likely for fish to eat such clay. Hence, it is still uncertain whether fish like flounders contained a high level of radioactive cesium because they dwell at the sea bottom.

2.3.4 *Forthcoming Challenges to Prevent the Spread of Radioactive Contamination in Food*

In sum, during the year 2011, the FDNPP accident had a large impact on the agricultural commodities. The accident has contaminated many agricultural products with radioactive cesium as shown in this chapter. However, after 1 or 2 years from the accident, the level of the radioactive cesium found in most of the test samples conducted for various agricultural products became lower than the safety standard. This indicates that the condition of the radioactive contamination in food after the Fukushima disaster is improving and the risk of food contaminated with radioactive materials to be sold in the market is becoming smaller.

On the other hand, we need to note that not all agricultural products are entirely free from radioactive contamination and, though small it may be, there is still a jeopardy of getting exposed to radiation by eating products like wild mushrooms, edible wild plants, and fish that are harvested near the FDNPP.

In this sense, it is important for the government to continue to conduct radiation tests for agricultural products and to implement strict policies to prevent food that contains radioactive materials from being distributed in the market.

References

Aruga, K. (2016). *Genpatsujiko to Fuhyou Higai (The nuclear accident and the reputation damage)*. Kyoto: Showado Inc.

Consumer Affairs Agency (CAA). (2014). Q & A for food and radiation (in Japanese). http://www.caa.go.jp/jisin/pdf/141113_food_qa_ver9.pdf Accessed February 4, 2017.

ICRP. (2007). The 2007 Recommendations of the International Commission on Radiological Protection. ICRP Publication 103. *Annals of ICRP* (Vol. 37, pp. 2–4).

International Atomic Energy Agency (IAEA). (2008). The international nuclear and radiological event scale user's manual 2008 edition. http://www-pub.iaea.org/MTCD/Publications/PDF/INES2013web.pdf. Accessed January 20, 2017.

Ishikawa, N. (2014). *Koushou Fukushima Genshiryokujiko (Investigation study of the Fukushima nuclear accident) (in Japanese)*. Tokyo: The Japan Electric Association.

Ministry of Health, Labour and Welfare (MHLW). (2012). Shokuhin chu no houshaseibushitsu no aratana kijunchi ni tuite (About the new standards for radioactive materials contained in food) (in Japanese). http://www.mhlw.go.jp/topics/bukyoku/iyaku/syoku-anzen/iken/dl/120117-1-03-01.pdf. Accessed February 2, 2017.

Nanayama, F., Saito, Y., Aziz, S., et al. (2007). Preliminary reports about tsunami traces and disasters of the 17th July, 2006 Java tsunami (in Japanese). *Chishitsu News, 636,* 42–51.

National Institute for Land and Infrastructure Management (NILIM). (2005). Sumatra tou oki daijishin oyobi Indoyou tsunami higai ni kansuru kinkyu kenkyu (An urgent research on the great earthquake of Sumatra and the Indian Ocean tsunami disaster) (in Japanese). http://www.nilim.go.jp/lab/beg/foreign/kokusai/sumatera2005.pdf. Accessed January 20, 2017.

Nuclear Emergency Response Headquarters (NERH). (2011). Report of Japanese government to the IAEA ministerial conference on nuclear safety: the accident at TEPCO's Fukushima Nuclear Power Stations (in Japanese). http://japan.kantei.go.jp/kan/topics/201106/iaea_houkokusho_e.html. Accessed January 20, 2017.

Radiation Science Center (RSC). (2013). Houshasen no Mamechisiki: Kurashinonaka no Houshasen (Useful knowledge of radiation: radiation in our living) (in Japanese). http://rcwww.kek.jp/kurasi/kurashi-all.pdf. Accessed February 20, 2017.

Yamada, T. (2013). Mushrooms: radioactive contamination of widespread mushrooms in Japan. In T. Nakanishi & K. Tanoi (Eds.), *Agricultural implications of the Fukushyima nuclear accident* (pp. 163–176). Tokyo: Springer.

Chapter 3
What Is Reputation Damage?

The purpose of this book is to provide a useful resource for conducting policies to mitigate the effects of economic damage from rumors regarding the food produced near the Fukushima Daiichi Nuclear Power Plant (FDNPP). Such economic damage related to rumors is often called the reputation damage in Japan, but this term is not a properly defined academic term and is popularized by the media. The definition and use of reputation damage often vary among different publications.

Hence, in this chapter, I would like to explain how reputation damage is defined in this book. Then I will discuss the causes of reputation damage. In the first section, I will elucidate the meaning of reputation damage in this book. Then, in the second section of the chapter, I will show why reputation damage occurs in certain situations.

3.1 What Is Reputation Damage?

Reputation damage in this book is about the loss of economic profit due to the effect of the bad reputation in food produced near the FDNPP but first I would like to introduce the commonly used definition for the term reputation damage. According to Sekiya (2011, p. 12), who is the author of the book Fuhyou Higai (reputation damage), reputation damage is economic damage caused after people avoid buying, visiting, and trading certain products. Economic damage can occur when consumer beliefs shift from a positive mindset to a negative one due to many numbers of reasons; social problems (event, accident, environmental pollution, disaster, and stagnation) related to the products reported by the media. Reputation results from how people judge the quality or character of one thing and reputation can turn bad when people start believing rumors that lack concrete proof of any sort and is based on uncertain information.

Reputation damage occurs when people start reacting to such bad reputation without clear facts or truths and avoid buying products with a bad reputation.

K. Aruga, *Consumer Reaction, Food Production and the Fukushima Disaster*,
DOI 10.1007/978-3-319-59849-9_3

On the other hand, when the rumor that is influencing consumer's reaction to avoid buying products produced near the FDNPP is arisen from a sound reasoning, economic effects on the food producers from such behavior will not be recognized as the reputation damage.

Using this standard definition for reputation damage, I would like to explain the reputation damage related to food produced near the FDNPP, which is the primary focus of this book. In this book, the food produced in areas near the FDNPP will be affected by reputation damage if the rumor regarding food products of regions near FDNPP is not based on any fact and if the producers of such food products are losing profits due to this rumor.

Following this definition of reputation damage, I would like to provide the conditions where the loss related to rumors become a reputation damage in this book. The first condition is that the rumors and information related to the safety of food produced near the FDNPP are not supported by any evidence. The second condition is that the reaction of people against the rumors and information about the food produced near the FDNPP is not based on their decisions and that these people are avoiding such food products without verifying the authenticity of the rumors and information. Thus, when people's decision to avoid buying food produced near the FDNPP rely on their decision and stem from some evidence, we assume that this type of behavior is not a reputation damage.

Let us say a parent chooses not to purchase a particular product produced near the FDNPP because the parent has the knowledge that young people are more susceptible to radiation poisoning, and the parent is trying to protect their children. Economic consequences due to this type of individual's avoiding behaviour are not considered to be reputation damage. Such consumers are avoiding food produced near the FDNPP because their decision rely on some evidence and knowledge and their decision is not just the result of bad rumors. On the other hand, if a consumer's behavior of avoiding food produced near the FDNPP is because the consumer has no knowledge of radiation and is simply swallowing misinformation without verifying those, effects from this consumer is considered as reputation damage.

In the next section, I would like to discuss the causes of reputation damage and argue why there are consumers who avoid purchasing food produced near the FDNPP even when it meets the national food safety standards for radiation.

3.2 Causes of Reputation Damage

One of the reasons why reputation damage occurred in food products of regions near the FDNPP is related to psychological factors that affect people's willingness to buy. In this section, I would like to discuss the causes of reputation damage from an economics point of view because economics has long provided factors affecting consumer's willingness to buy in various conditions. When we consider the problem of reputation damage from an economics aspect, the causes of reputation damage can be explained by the following three factors. The first factor is that food products often have substitutes. The second factor is that information about a particular food product contains vague or uncertain information. The third factor is that there is an asymmetric information problem between the producers and consumers of food products. In this section, I would like to explain the causes of reputation damage from these three aspects.

3.2.1 *Food Products Have Substitutes*

Here, I will explain why the characteristics of food products having many substitutes lead to reputation damage. A substitute in economics is goods whose demand increases when the price of its substitutive good increases. For example, bread is considered a substitute for rice because if the price of rice increases, a shift of demand occurs from rice to bread and the demand of bread increases accordingly. These are goods that consumers buy to replace ones whose prices increases or become unavailable.

The reason why the price of food produced near the FDNPP to decline after the Fukushima disaster is because most of these food products have substitutes that are produced in regions further apart from the FDNPP. This kind of shift from food produced near the FDNPP to those produced further away from the FDNPP is more likely to occur when there is not much difference between the food products of regions near the FDNPP and their substitutes.

When rumors that the food products of regions near the FDNPP have the risk of radioactive contamination spread out, many consumers started to buy substitutive products from regions away from the FDNPP. As most of the food products of regions near the FDNPP were very similar to food of regions away from the FDNPP, many consumers shifted their demand for substitutes to avoid any risk involved in the products of regions near the FDNPP. Hence, the food products of regions near the FDNPP were very susceptible to bad rumors, and this is why the producers of these regions received economic damages from losing reputation.

What makes it tough to conduct policies to cope with reputation damage in food products is that there are people who are indifferent with the truthfulness of the rumor. These people often do not want to spend their time to verify the authenticity of the reputation even if the rumor is wrong. Thus, establishing a strict regulation to strengthen the food safety level for the product produced near the FDNPP will not help increase the demand of these people because these people avoid any product of regions near the FDNPP as long as bad reputations flow in the product.

3.2.2 *Uncertain Information*

Next, I would like to discuss the uncertainty in the information regarding the food produced near the regions of FDNPP. As already mentioned before, having an uncertain information is one of the causes of reputation damage. First, using the definition of risk and uncertainty of Costanza et al. (1997), I would like to clarify the meaning of uncertainty.

A risk is an event where its probability of occurrence can be statistically estimated from experiences of similar events that occurred in the past (Costanza et al. 1997). For example, the risk of causing an accident from driving a car is a problem of risk because the probability of car accidents and type of accidents that might occur can be somewhat estimated from the historical data of accidents that took place in the past. This kind of incident whose degree of threat can be statistically determined is considered as the problem of risk.

On the other hand, real uncertainty is an event where the probability of occurrence is unknown at the current state. For instance, the problem of sea level rise due to climate change is an incident whose likelihood of occurrence and the degree of threat cannot be estimated because it is an incident we have not experienced and there is no reliable data available from the past. Such problem whose level of risk is not predictable from the experience is the real uncertainty problem.

Uncertain information is a case where the information contains this real uncertainty problem and is difficult to statistically identify its reliability.

Let us consider this uncertain information problem to the case of radioactive contamination of food after the Fukushima disaster. A nuclear accident that led the agricultural products to become contaminated with radioactive material is a disaster we have never experienced so the problem is full of information uncertainties and information on the threat of radioactive contamination can only be transferred to the consumer with uncertainties. This is one of the reasons why there are many consumers who are avoiding to buy food produced near the FDNPP even after the Japanese government set strict safety standards of radioactive materials for food products. When there persists uncertain information in the food product it is tough for consumers to figure out whether the product is safe or not and this is why rumors can cause economic damage to the food producers of regions near the FDNPP.

However, if we can gradually remove the uncertainties in the food produced near the FDNPP with the help of transmitting reliable information by food specialists and scientists, we might be able to increase the credibility of the information. If so, more people might start to trust the national standards of radioactive materials for food products. Hence, I believe conducting policies to develop technologies for estimating radioactivity in food products and announcing credible information will help mitigate the effects of the reputation damage.

3.2.3 Asymmetric Information Problem

In this subsection, I will describe the asymmetric information problem between the producers and consumers of agricultural commodities. Asymmetric information problem is the third factor of the causes of reputation damage introduced in this book.

Asymmetric information problem arises when differences in the quality and volume of information exist between the producers and consumers. Regarding the information on the safety of the food produced near the FDNPP, the producers of the food know more about how and when the food was produced but consumers have little information about it, and hence there is an information asymmetry between them. For example, the producers have information on what type of soils and fertilizers were used for growing the product while the consumers have no information about this matter. Like this example, consumers have less information than the producers about the food product. When consumers only have poor information it is challenging for them to verify whether the rumors on food produced near the FDNPP is a falsehood or not. Hence, it is important to build an environment where the consumers can obtain more information about the product such as that of the production method and the producer.

Most grocery stores only provide the consumers with information about the origin of the product and expiration or production dates but such information is not enough for the customers to decide whether the food they buy is safe or not. Therefore, most consumer's decision to avoid products of regions near the FDNPP is only based on the information related to the origin of the product. Reputation damage occurs because when consumers have little information, they are more likely to be affected by rumors because trying to search the safety of the food produced near the FDNPP by themselves is not an easy thing to do.

In most cases, consumers have little information about the products of regions near the FDNPP and they do not have many incentives to gather and learn information related to the products. Some of them perhaps even do not know that for selling food products the products have to meet the strict national food safety standards for radiation. Many consumers have little knowledge in radiation and do not know that radiation exists in nature and only becomes harmful when the amount exceeds a certain level. If there are such consumers, it is important for the government to fill the information asymmetry between the producers and consumers and educate the consumers to have more knowledge in radiation. It is believable that such effort will help reduce the economic damage from bad reputation on the food produced near the FDNPP.

3.2.4 To Overcome Asymmetric Information

Finally, from an economics point of view, I would like to explain what policies will be effective for mitigating the effects of asymmetric information.

The first effective action to fix the asymmetric information problem is related to the economics term signaling. Signaling is an action where an economic player holding more information supplies information for the player with less information to fill the information gap between them. For example, putting a label that certifies radiation safety on the products of regions near the FDNPP can be an effective way for the producer to transmit information to the consumer about the safety of the products. To do this, the food producers of the regions near the FDNPP can measure the radiation levels of soils and fertilizers used for the production and create a special label to show the consumers that the product is safe from radiation. Another action the producers can perform to fill the information gap is to invite the consumers to the farmlands and directly explain how safe their products are.

The second effective method for filling the gap between the producers and consumers is related to the action of the consumer, where the consumers seek a way to obtain information from the producers. In economics, this type of action where the player having less information strives to gain more information from the informationally superior player is called the screening. For example, if the consumers can use radiation measuring instruments at a grocery store they can get information about the safety of the food product by themselves. If there is a way for the consumers to acquire more information about the safety of food produced near the FDNPP by themselves, it is believable that the effects of asymmetric information can be mitigated.

After going over the causes of reputation damage in this chapter, it seems that the main cause of the reputation damage is attributed to whether or not the consumers can trust the safety of the food produced near the FDNPP. Transmitting reliable information about particular products is essential for producers of regions near the FDNPP to establish trust with its consumers. Hence, to alleviate the reputation damage, it is important for the government to conduct policies to remove uncertain and asymmetric information problems. I am hoping the investigation of this book based on the questionnaire performed on the Japanese consumers would help understand what kind of information is needed to regain the consumer's trust on foods produced near the FDNPP.

References

Costanza, R., Cumberland, J., Goodland, R., & Norgaard, R. (1997). *An Introduction to Ecological Economics*. Florida: St. Lucie Press.

Sekiya, N. (2011). *Fuhyou Higai (Reputation damage) (in Japanese)*. Tokyo: Kobunsha Inc.

Chapter 4
Conditions of the Agricultural Commodity Markets Before and After the Fukushima Disaster

In this chapter, we investigate the conditions of the agricultural commodity markets of regions near the Fukushima Daiichi Nuclear Power Plant (FDNPP) before and after 2011, the year the Fukushima disaster took place.

This book examines the consumer reaction toward agricultural products from regions near the FDNPP using a questionnaire conducted with 8700 people throughout Japan. In the questionnaire, we asked people to respond to the following nine products: rice, cucumbers, apples, *shiitake* mushrooms, beef, pork, eggs, tuna fish, and *wakame* seaweed. Therefore, we look into the conditions of the markets for these nine agricultural products. For this purpose, I will use the data for the price and transaction value of the nine products and find out how the agricultural markets changed before and after the Fukushima disaster.

4.1 Why Do Agricultural Commodity Prices Change After the Fukushima Disaster?

Before introducing the data for the agricultural commodity markets of regions near the FDNPP, I would like to explain cases where the price of agricultural products increase and decrease as price and transaction value change.

First, let us consider what condition prices of agricultural products from regions near the FDNPP increase after the Fukushima disaster. After the FDNPP accident, agricultural production became impossible in some areas near the FDNPP because of the effect of the tsunami and radioactive contamination. In such regions, the volumes of production will plummet as areas that cannot harvest agricultural products increase. If we explain this situation regarding supply and demand relationships, supply decreases assuming that demand remains constant, the price will increase due to a relative increase in the demand. However, it is hard to determine how this price increase affects the transaction value of a product. Transaction value

K. Aruga, *Consumer Reaction, Food Production and the Fukushima Disaster*,
DOI 10.1007/978-3-319-59849-9_4

is the product of price and volume of trade so it will go up when a level of the price increase is large enough to cancel out the effects of the decline in production. However, transaction value will go down if the level of the price increase is not significant enough to counteract with the decreased production.

Second, when will the price of agricultural products of regions near the FDNPP drop after the Fukushima disaster? As explained in the previous chapter there are many substitutes for agricultural products. Therefore, it is possible to think that if consumers start to become anxious about the risk of radioactive contamination of the products produced near the FDNPP, many people will begin buying agricultural products from regions apart from the FDNPP as substitutes of the goods produced near the FDNPP. If many consumers are avoiding agricultural products from regions near the FDNPP, then, these products will mostly likely remain unsold, causing their prices to drop. This is the probable cause for prices of agricultural products of regions near the FDNPP to drop after the Fukushima disaster.

When prices of products of regions near the FDNPP plunge as demand shifts to their substitutes, it can be expected that their transaction values will drop accordingly. This is quite obvious because transaction value is the product of price and demand, and decrease in price and demand will lead to the drop in transaction value.

Hence, it is likely that when bad reputation about the products of regions near the FDNPP spreads out, more consumers will avoid buying products of these regions and this will cause the price and transaction value of these products to decrease. If we find the price and transaction value of agricultural products of regions near the FDNPP are declining, it will imply that the consumer's willingness to buy these products is decreasing.

In the following section, considering these possibilities of changes that may occur in the agricultural commodity markets of the regions near the FDNPP, I would like to investigate the conditions of these markets before and after the Fukushima accident.

4.2 Conditions of the Agricultural Commodity Markets Before and After 2011

4.2.1 Market Price and Value of Production for Rice

Figure 4.1 illustrates the price and value of production of rice produced in Fukushima, Ibaraki, Tochigi, and Yamagata prefectures during 2009–2013. The vertical axis of the price figure represents the annual average price of 60 kg of brown rice and that for the value of production figure denotes the total annual value of production in 100 million yen. The figure is created based on the data obtained from the homepage of the Ministry of Agriculture, Forestry and Fisheries of Japan (see Aruga 2016 for details).

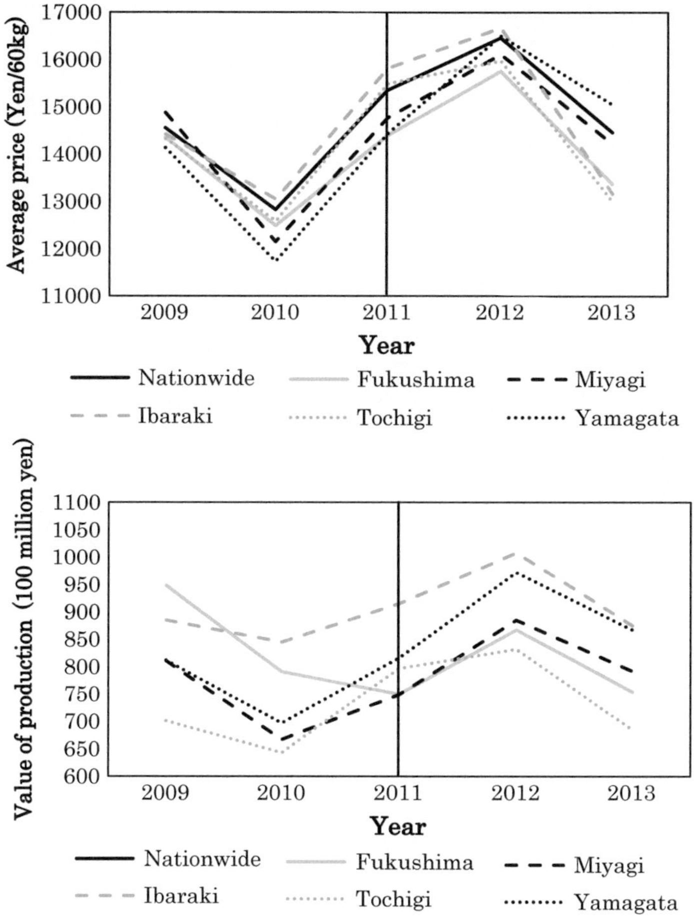

Fig. 4.1 Market price and value of production for rice. *Source* Aruga (2016, p. 34)

It can be presumed from the figure that the rice from Fukushima and Tochigi prefectures had an economic impact from the nuclear accident. You can see from the figure that the prices of Fukushima and Tochigi rice were ranked the third and second highest among the five prefectures in 2010 but fell to fifth and fourth in 2012 indicating that rice prices for these prefectures became relatively cheap after the nuclear accident. The value of production of the Fukushima and Tochigi rice also had a similar declining trend in their ranks in 2012. In particular, the value of production of the Fukushima rice was ranked second in 2011, but it became the fourth in 2012, which suggests that the effects from the nuclear accident have been very dramatic for the Fukushima rice market. Although the levels of the price and value of production for the Fukushima and Tochigi rice comparatively became lower than those of other three prefectures after 2011, you can see from Fig. 4.1

that both the price and value of production increased from their pre-accident values, implying that there were quite a few people who did not change their demand for the Fukushima and Tochigi rice even after 2011. Hence, it is likely that the effect of the Fukushima disaster on the rice market of regions near the FDNPP was not significant enough for their price and value of production to become lower than the pre-accident level.

4.2.2 Market Price and Transaction Value of Cucumbers

In this subsection, we will investigate the conditions of the cucumber markets of regions near the FDNPP. Figure 4.2 represents the annual average wholesale price and transaction value of cucumbers produced in Fukushima, Miyagi, Ibaraki, Tochigi, and Yamagata prefectures that are traded at the Ota Metropolitan Central Wholesale Market in Tokyo (see Aruga 2016 for details). The price in the figure is the average annual prices of 1 kg of cucumbers in Japanese yen, and the transaction value represents the total amount of transaction of cucumbers from the five prefectures whose units are in 100 million yen.

First you can see from Fig. 4.2 that the national average price of cucumbers at the Ota Tokyo market dropped a little in 2012 but recovered and increased in 2013. However, most of the prices of cucumbers harvested in the regions near FDNPP show a huge plunge in 2012. Among the cucumber prices of the five prefectures illustrated in the figure, all prefectures except for Ibaraki show a drop in their prices after the Fukushima accident of 2011.

Second, you can find that the transaction values also decreased in 2012 in all prefectures except for Yamagata Prefecture. It is apparent from the figure that the depreciation of the Fukushima cucumber transaction value was drastic during the 2010–2012 period. The transaction level of Fukushima cucumbers was 1.25 billion yen in 2010 but became 910 million yen in 2012 yen, a 30% plunge from the level of 2010. The Ibaraki cucumbers also had a drop in its transaction value while its price was stable during the 2010–2012 period: the price decreased from 550 million yen in 2010 to 500 million yen in 2012, about a 10% reduction from the pre-accident level.

The analysis of the changes in the price and transaction value of cucumbers during the 2010–2012 period tells us that the cucumber markets of regions near the FDNPP received a significant impact from the Fukushima disaster. Both the price and transaction value decreased in the cucumber markets of regions near the FDNPP after the Fukushima disaster. It is likely that consumer's avoiding behavior toward the cucumbers of regions near the FDNPP is one of the causes of this decline after 2011.

None of the test samples of the cucumbers produced in Fukushima Prefecture contained radioactive cesium above the national safety standards of radioactive cesium, so even the Fukushima cucumbers was safe from radiation based on the national standards. However, it is believable that not all consumers believed this information or tried to search the authenticity of this information. Thus, it is

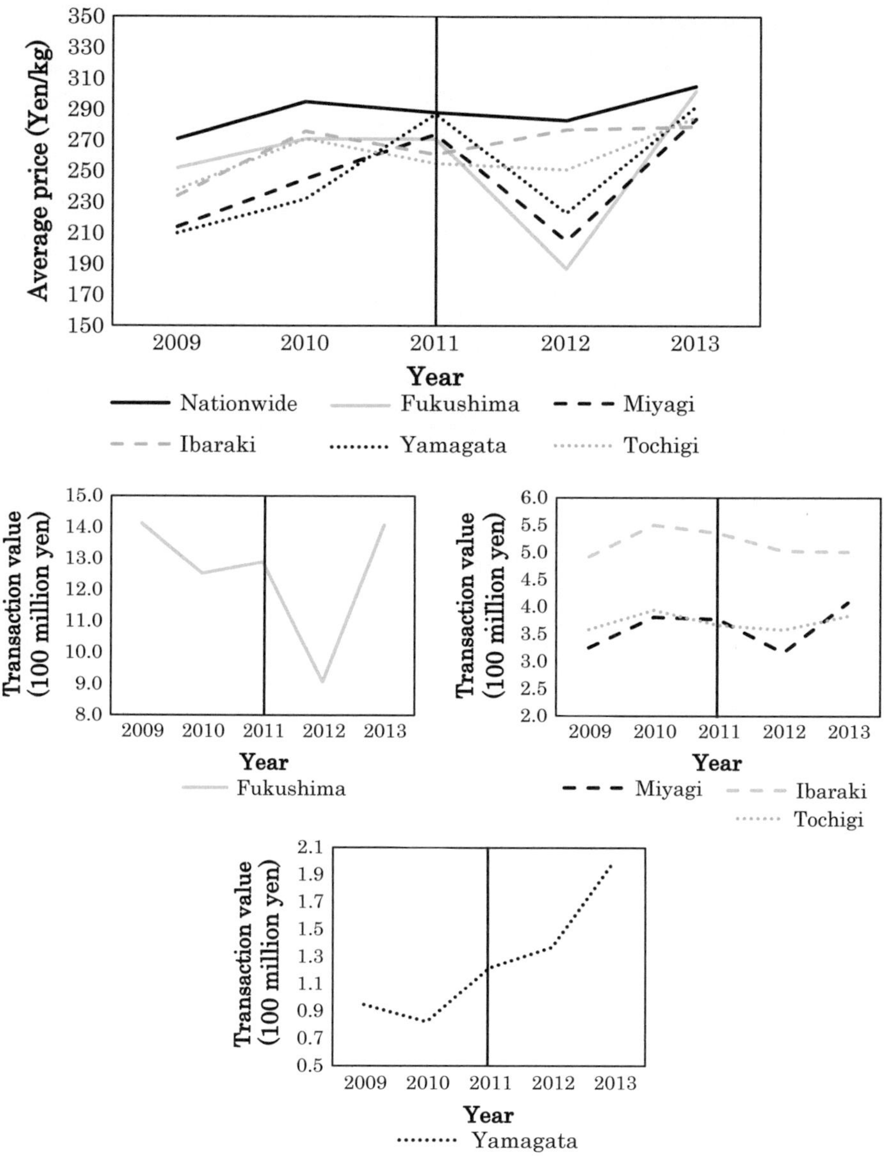

Fig. 4.2 Market price and transaction value of cucumbers. *Source* Aruga (2016, p. 36)

probable that some of the consumers are avoiding the cucumbers of regions near the FDNPP just because the regions are located near the FDNPP. It is likely that such consumer behaviors decreased the price and transaction value of the cucumbers of these regions. Hence, it is possible to say that reputation damage did take place after the Fukushima accident in the cucumber markets of regions near the FDNPP.

4.2.3 Market Price and Transaction Value of Apple

Figure 4.3 depicts the average annual price and transaction value of apples produced in Fukushima, Yamagata, Aomori, Nagano prefectures that are traded at the

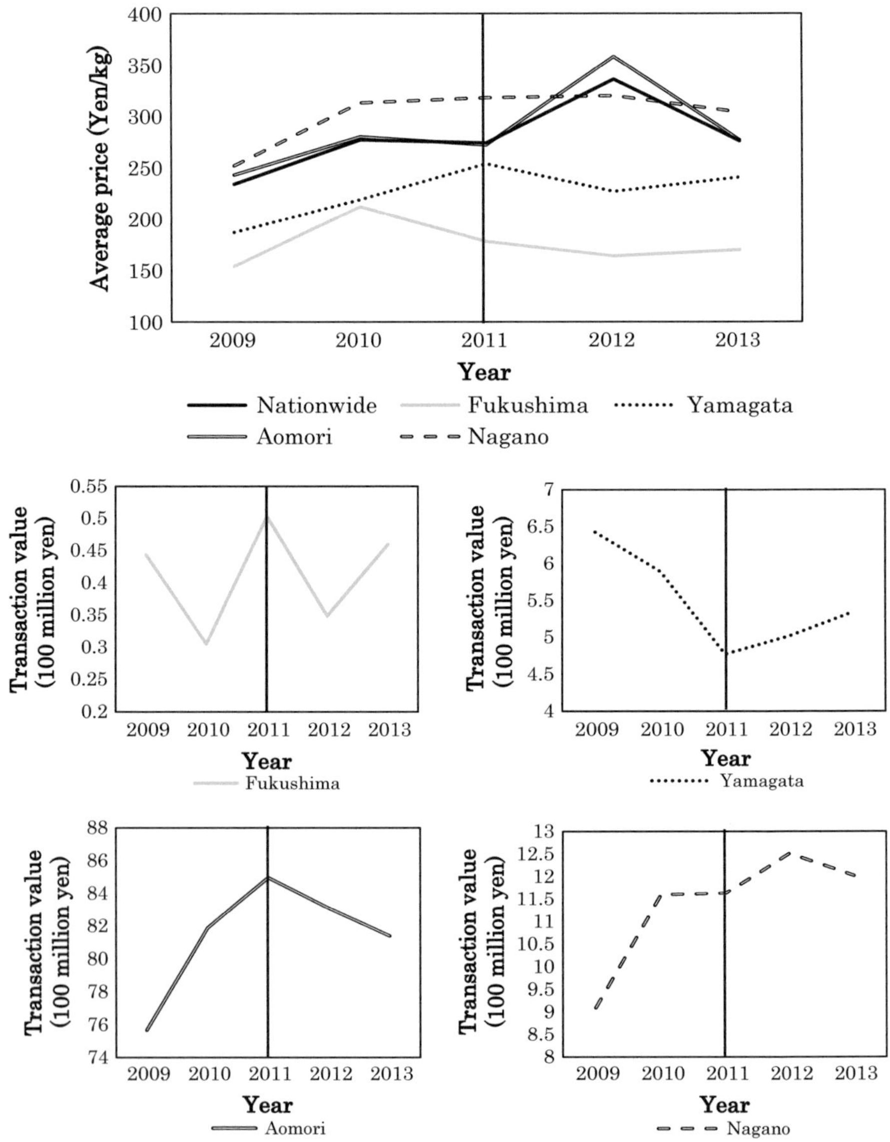

Fig. 4.3 Market price and transaction value for apples. *Source* Aruga (2016, p. 38)

Ota Tokyo market.[1] The vertical axis of the price represents the price of per kilogram of apple in yen, and that of the transaction value is 100 million yen.

Comparing the 2010 and 2012 average annual prices of apples in the figure, you can see that except for the apples of the Fukushima Prefecture the prices of apples have increased after the Fukushima accident. However, the price of Fukushima apples plummeted from 212 to 164 yen/kg in 2012. This is more than a 20% depreciation from the level of 2010. For the transaction value, most of the prefectures increased from 2010 to 2012, but that of the Yamagata Prefecture decreased from 2010 to 2012.

As the price of Fukushima apples dropped after the Fukushima disaster, it can be expected that the transaction value will decline after the catastrophe but the transaction value did not change much compared to the level of 2010. However, comparing the transaction value of the 2011 and 2012 Fukushima apples, the transaction value plunged dramatically, contrasting to the changes in the transaction value of the Nagano apples which increased dramatically from 2011 to 2012. This might be because the consumers shifted their demand from the Fukushima to Nagano apples to avoid the apples from the Fukushima Prefecture.

For apples, none of the test samples were found to be contaminated with radioactive cesium so it is likely that the price decrease of the Fukushima apples is related to the drop in the transaction value during the 2011–2012 period. It is believable that this change in the price is related to the harsh rumors toward the Fukushima apples and the Fukushima apple market did receive economic damages from the reputation.

4.2.4 Market Price and Transaction Value for Shiitake Mushrooms

Figure 4.4 represents the change in the price and transaction value of the forestry product shiitake mushrooms for the 2009–2013 period. The price is the annual average yen per 1 kg of raw shiitake mushroom. The transaction value is the annual value of shiitake mushrooms traded at the Ota Tokyo market, and its unit is 1 million yen. The figure displays the price and transaction values of shiitake mushrooms harvested in Fukushima, Miyagi, Ibaraki, Tochigi, and Yamagata prefectures.

Among the five prefectures in Fig. 4.4, four prefectures except the Yamagata Prefecture have areas that are restricted to ship their wood mushrooms because some of the mushrooms from these regions were contaminated with radioactive materials after the accident. Such shipping restriction for the wood mushrooms in the four prefectures remains as of 2015. However, shitake mushrooms that are

[1]Apples traded in the Ota market consists of 13 different kinds of apples such as Tsugaru, Akane, Jonah Gold, Ruby apples, Fuji, and Sekaiichi. The price and transaction value of the apple for Fig. 4.3 denotes the average values when all these 13 types of apples are combined into one data.

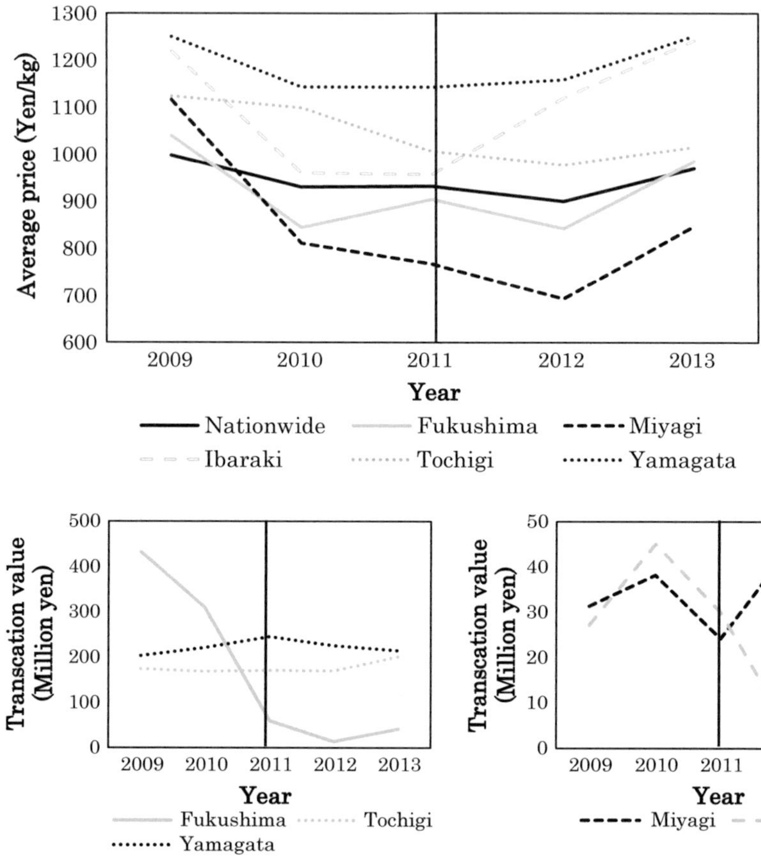

Fig. 4.4 Market price and transaction value for shiitake mushrooms. *Source* Aruga (2016, p. 40)

cultivated with sawdust block are not restricted in any of the five prefectures as long as the sawdust is not produced with contaminated logs.

It can be expected that the price and transaction value are chiefly affected from the Fukushima disaster because wood mushrooms of some parts of the regions near the FDNPP had shipping restrictions. However, most of the raw mushrooms sold in the market are now grown with sawdust blocks. Therefore, the price and transaction value taken from the Ota Tokyo market mostly reflects the price of sawdust grown mushrooms. Indeed, according to the Forestry Agency (2015), about 90% of the raw mushrooms produced in Japan is sawdust grown mushrooms. Therefore, the effect of the shipping restriction might be small on the raw mushrooms market. So next, I would like to look at what happened to the price and transaction of the shiitake mushroom markets after the Fukushima disaster.

First, comparing the price of the shiitake mushrooms of 2010 and 2012, the price for most of the prefectures stayed at the same level or dropped beside the Ibaraki Prefecture (see Fig. 4.4). Second, it is apparent from the figure that the transaction values of Fukushima and Ibaraki mushrooms have largely depreciated after the Fukushima accident. This result is interesting because although Tochigi and Miyagi prefectures also had areas where shipping wood mushrooms became restricted it was the Fukushima and Ibaraki prefectures that had a large impact on their mushroom markets after the accident. As some parts of Tochigi and Miyagi prefectures are closer to the FDNPP than parts of Fukushima and Ibaraki prefectures, it is strange that their impacts are so different.

The reason for this difference is perhaps due to the lack of information on the level of radioactive contamination for mushrooms. If the consumer knew that the Tochigi and Miyagi prefectures too had areas whose wood mushrooms became highly contaminated with radioactive materials, it could be that the price and transaction value of these prefectures also had a massive plunge similar to that of Fukushima and Ibaraki prefectures.

In summary, as seen in the Fukushima and Ibaraki markets, it is likely that the producers of the shiitake mushrooms of regions near the FDNPP did receive economic damage from the reputation and that this reputation damage is related to the lack of information about the condition of radioactive contamination for the shiitake mushrooms (Fig. 4.4).

4.2.5 Market Price and Transaction Value for Beef

Figure 4.5 shows the price and transaction value of Japanese beef produced in Fukushima, Miyagi, Ibaraki, Tochigi, Gunma, and Yamagata prefectures. The data is taken from the Shibaura Metropolitan Central Wholesale Market in Tokyo. The price represents the 1 kg of beef in yen, and the transaction value denotes the total value of beef transacted in the Shibaura market in 100 million yen.

Comparing the levels of 2010 and 2012 beef prices, the prices of beef for 2012 decreased in all five prefectures but you can see from the figure that most of the beef prices recovered to the level before the Fukushima accident in 2013 (Fig. 4.5).

The transaction values for the five prefectures had different movements. The transaction values of Fukushima, Miyagi, and Tochigi beef were stagnant after the accident and did not recover to the pre-accident level even in 2013. On the other hand, those for Gunma, Ibaraki, and Yamagata already recovered to the pre-accident level in 2012 and continued to increase in 2013. The reason why the transaction values of beef for Fukushima, Miyagi, and Tochigi hovered around the pre-accident level compared to those for Gunma, Ibaraki, and Yamagata is perhaps because some of the beef produced in these prefectures became contaminated with radioactive materials and some of the livestock farmers had to restrict their shipping. It is known that the beef from the Fukushima, Miyagi, and Tochigi became contaminated with radioactive cesium after the nuclear disaster because the livestock have eaten some rice straw contaminated with radioactive materials.

The Fukushima beef market seemed to have the largest impact from the Fukushima accident because even after none of the test samples for beef contained

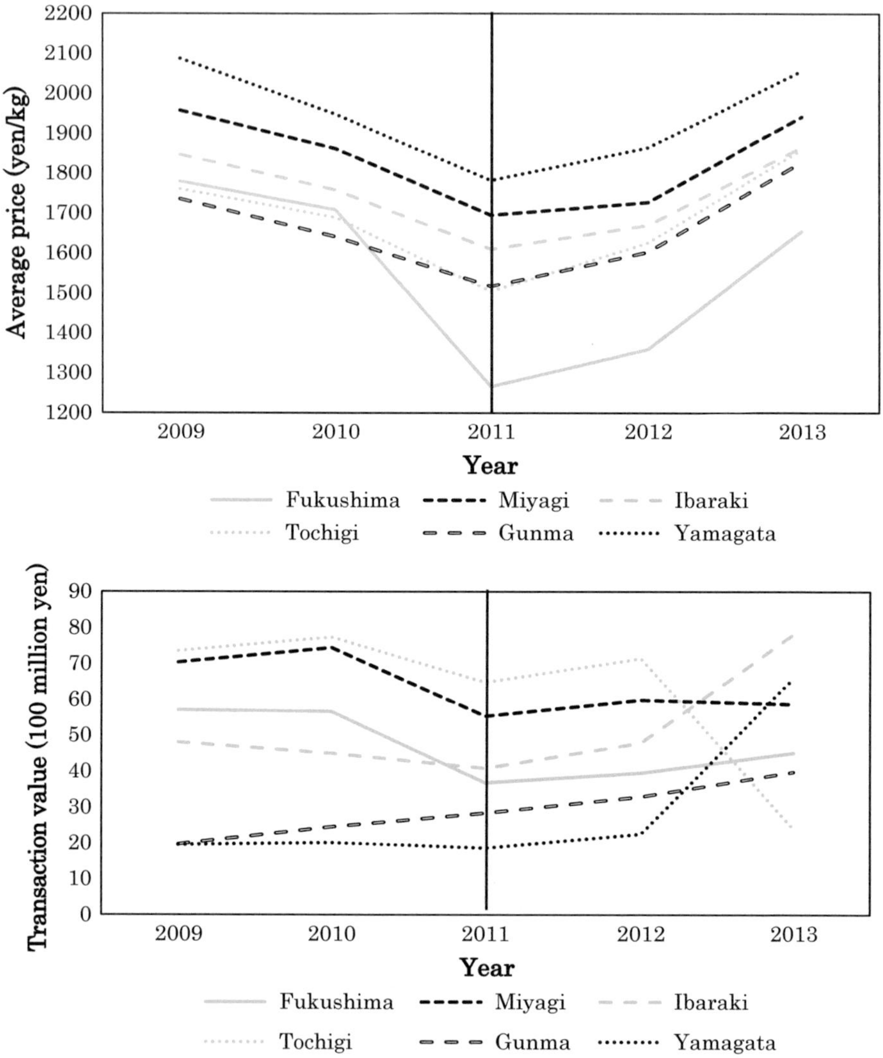

Fig. 4.5 Market price and transaction value for beef. *Source* Aruga (2016, p. 42)

radioactive cesium above the national food safety standards in 2013 (see Table 2.7), the average beef price of Fukushima did not recover to the pre-accident level. This result is different compared to the price movements of Miyagi and Tochigi prefectures because the prices of beef for these prefectures at least recovered in 2013. This difference in the speed of price recovery after the accident might be because the Fukushima beef had a reputation damage just because the name Fukushima reminded the consumers of FDNPP.

4.2.6 Market Price and Transaction Value for Pork

Figure 4.6 is also created from the data obtained from the Shibaura Tokyo meat market. The figure shows the price (yen/kg) and transaction values (100 million yen) of pork from the Fukushima, Ibaraki, Tochigi, Gunma, and Iwate prefectures.

Comparing the 2010 and 2012 average pork prices, prices of four prefectures except the Fukushima Prefecture stayed at the same level or decreased in 2012 and price of Fukushima too dropped from 2011 to 2012.

For the transaction values of pork, the levels remained stagnant until 2012 besides the Ibaraki pork, which indicates that the pork markets of regions near the FDNPP received large impact from the Fukushima accident. However, most of the pork markets of regions near the FDNPP in the figure recovered in 2013 and the recovery of the pork market seemed to be quicker than the beef market.

The possible reason for the pork market restored more quickly compared to the beef market is perhaps the economic damage for pork was smaller than that for

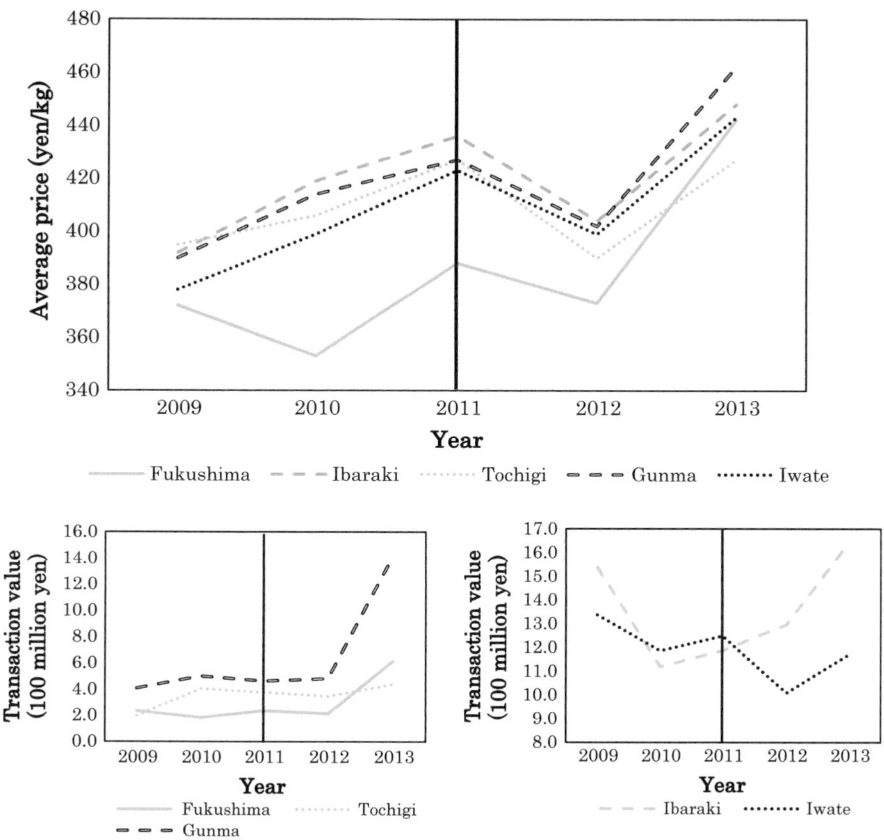

Fig. 4.6 Market price and transaction value for pork. *Source* Aruga (2016, p. 45)

beef because most of the pig feeds are imported outside of Japan (AMSJ 2008, pp. 140–141) and consumers felt safe about the risk of radioactive contamination for pork.

4.2.7 *Market Price and Transaction Value for Chicken Eggs*

In this subsection, I will explain the conditions of the chicken eggs produced in regions near the FDNPP using Fig. 4.7. This figure is created from the data

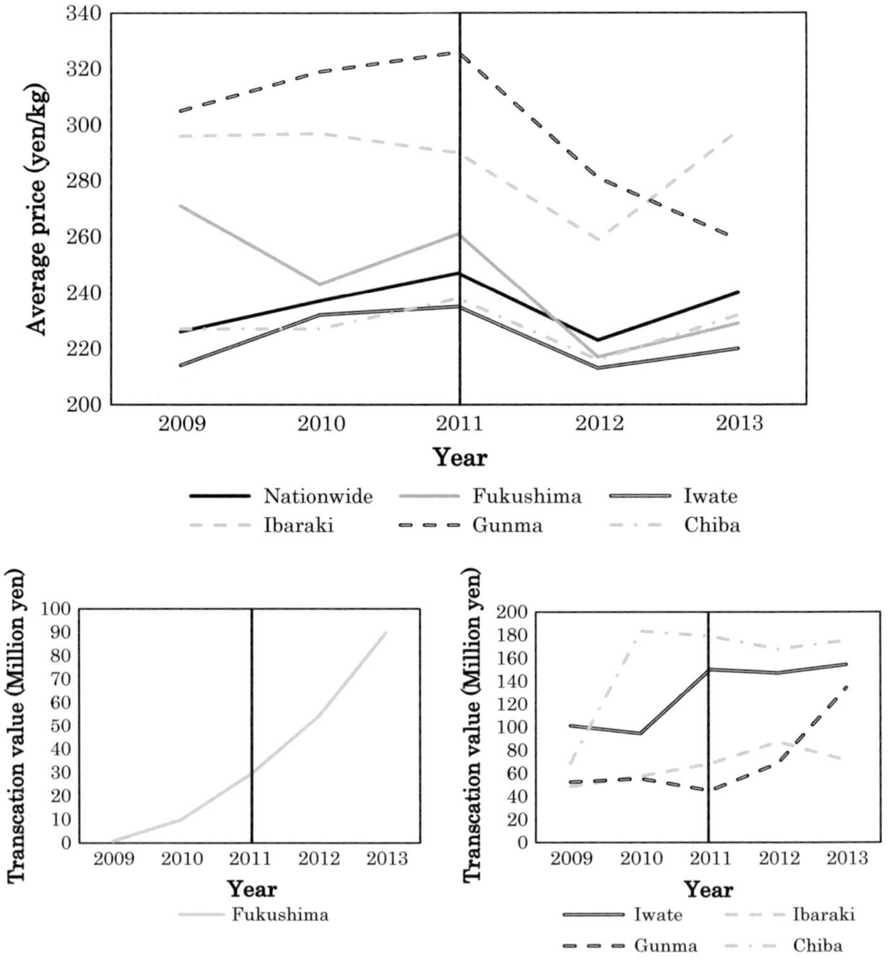

Fig. 4.7 Market price and transaction value for chicken eggs. *Source* Aruga (2016, p. 45)

obtained from the Tokyo Tsukiji wholesale market. The figure represents the price (yen/kg) and transaction value (million yen) of chicken eggs produced in Fukushima, Iwate, Ibaraki, Gunma, and Chiba prefectures.

From Fig. 4.7, the average chicken egg prices of five prefectures decreased from 2010 to 2012. For the transaction values, you can see from the figure that except for the Chiba chicken eggs they continued to increase from 2010 to 2012 even after 2011. Thus, although the prices dropped, the total transaction values were increasing so the decline in the prices is not probably related to the depreciation of consumer demand after the Fukushima accident. This tells us that the effect of the Fukushima disaster on the chicken egg markets was relatively smaller compared to the other agricultural markets we discussed in this chapter so far.

It is believable that the reason for the chicken egg markets of regions near the FDNPP to have small influence from the nuclear accident is because chickens are mostly fed with imported feeds and the risk of chicken eggs to be contaminated with radioactive materials are considered as small among consumers.

4.2.8 Market Price and Transaction Value for Tuna Fish

Here and in the next subsections, I would like to explain the condition of the seafood markets after the Fukushima disaster by investigating the conditions of the tuna fish and wakame markets.

In this subsection, we will look into the condition of the tuna fish market using the Tokyo Tsukiji wholesale market data. I used the annual average tuna fish price (kg/yen) and transaction volume (million yen) of Miyagi, Aomori, Tokyo, and Iwate prefectures. The data for the Fukushima prefecture was not available in the Tsukiji market data so we consider the data of the Miyagi Prefecture as the market data of region nearest to the FDNPP.

From Fig. 4.8, we can see that the average annual tuna fish prices of 2012 became higher than those of 2010 in three prefectures besides Aomori Prefecture. The transaction value of the Miyagi and Tokyo tuna fish also increased after the Fukushima accident suggesting a likelihood that the tuna fish market did not have much influence from the accident.

However, seeing the movements of the price and transaction value of Aomori tuna fish market both of them dropped from 2010 to 2012. Aomori has regions famous for their tuna fish like Oma whose bluefin tuna is a well-known brand in Japan and the volume and value of the transaction for tuna fish in Aomori Prefecture remain relatively large compared to other prefectures. Thus, though small it may be, the plunge in the Aomori tuna fish market after the Fukushima disaster is likely to be the effect of the nuclear accident and the change in the consumer reaction toward its tuna fish.

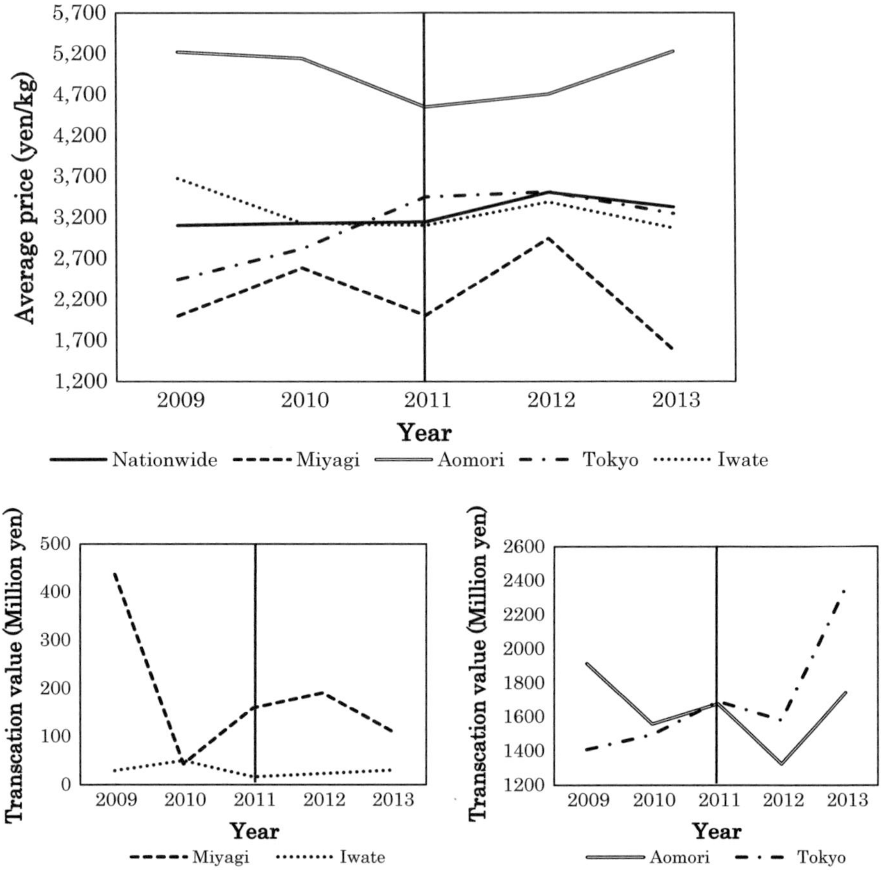

Fig. 4.8 Market price and transaction value for tuna fish. *Source* Aruga (2016, p. 47)

4.2.9 Market Price and Transaction Value for Wakame Seaweed

Here I would like to explain the condition of raw wakame seaweed markets of regions near the FDNPP. The data source here is also based on the Tokyo Tsukiji wholesale market, and we used the price (kg/yen) and transaction value (million yen) of wakame seaweed for Miyagi, Iwate, Tokyo, and Chiba prefectures.

Iwate Prefecture is famous for its Sanriku wakame brands but as seen in Fig. 4.9 the price of Iwate wakame seaweed plunged after the Fukushima accident. Comparing the prices of Iwate wakame seaweed of 2010 and 2012 in the figure, the Iwate price dropped from 585 to 456 yen/kg. The transaction value of the Iwate wakame also plummeted after the Fukushima accident so both the price and transaction value declined for the Iwate wakame seaweed market. This result might be telling us that the Iwate seaweed market is having a serious effect from reputation damage after the Fukushima accident.

The Miyagi wakame seaweed market also had a declining trend. The price of Miyagi wakame seaweed did not recover much in 2012, and its transaction value has been stagnant after the nuclear accident. It is believable that the Miyagi wakame seaweed too had a negative economic influence from the Fukushima incident.

On the other hand, as seen in Fig. 4.9, both the prices and transaction values of Tokyo and Chiba wakame seaweeds have increasing trends. These changes in the Tokyo and Chiba wakame seaweed markets are perhaps relevant to the shift in the consumer demand for Iwate and Miyagi wakame seaweed. It is likely that the consumers started to buy the Tokyo and Chiba seaweeds as substitutes of Iwate and Miyagi seaweeds (Fig. 4.9).

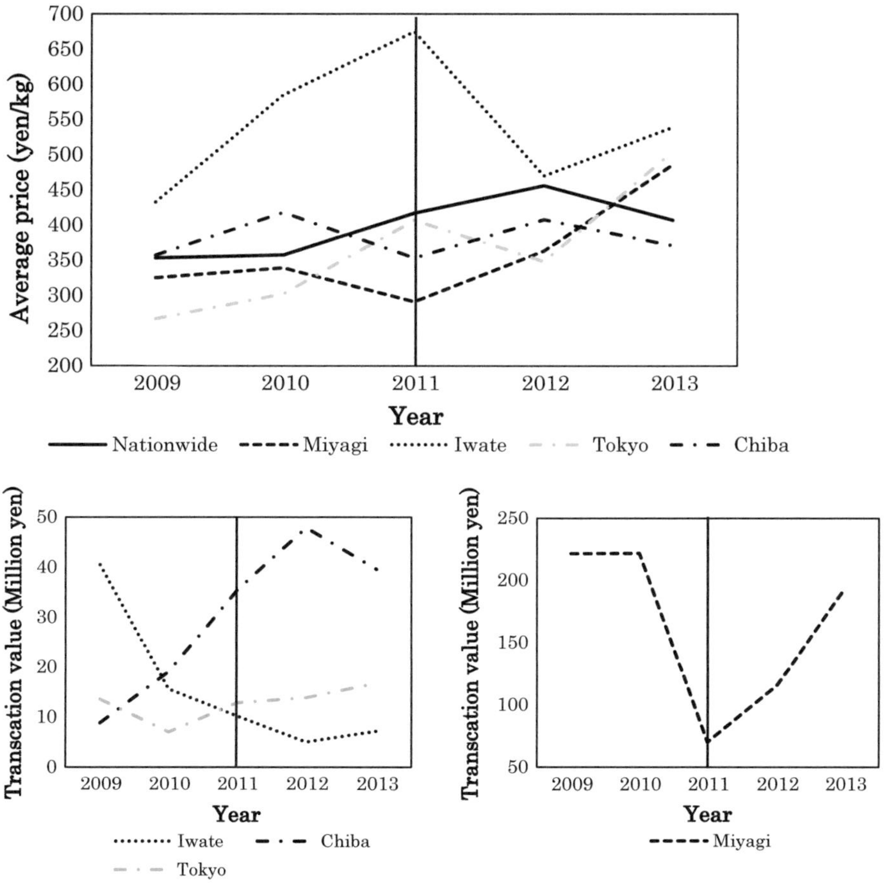

Fig. 4.9 Market price and transaction value for wakame seaweed. *Source* Aruga (2016, p. 48)

4.3 Influence of the Price Change on the Producers of Agricultural Commodities Near the FDNPP

From the above investigation on how the agricultural markets of regions near the FDNPP moved before and after the Fukushima accident, we found that all nine agricultural markets we analyzed have been somewhat affected from the accident. The price and transaction value of the nine agricultural products all had great changes in 2012, the year just after the Fukushima accident. These results indicate that the Fukushima accident caused a significant impact on the agricultural markets of regions near the FDNPP.

The decrease in price and transaction value of agricultural products of regions near the FDNPP in 2012 means that the Fukushima accident brought great economic damage to these regions. This plunge in the price and transaction value of agricultural products is likely related to the decline in the consumer's willingness to buy products from regions near the FDNPP. It is believable that this drop in consumer's willingness to buy is related to false rumors like every food products of regions near the FDNPP is contaminated with radioactive materials. However, as we have seen in Chap. 2, some agricultural products of regions near the FDNPP were never identified as contaminated. Then why do consumers believe such false rumors?

To find out this reason and to understand the factors behind the consumer reactions toward food produced near the FDNPP, in the next chapters, we will investigate the causes of reputation damage using the survey performed on consumers of all parts of Japan.

References

Agricultural Marketing Society of Japan. (2008). *Shokuryou Nousanbutsu no Ryutsu to Shijyou II (Distribution and markets of food and agricultural products II)*. Tsukuba: Tsukubashobo. (in Japanese).

Aruga, K. (2016). *Genpatsujiko to Fuhyou Higai (The nuclear accident and the reputation damage)*. Kyoto: Showado Inc. (in Japanese).

Forestry Agency. (2015). *Heisei 26 nen no shuyou na tokuyou rinsanbutsu no seisandoukou (Production status of the non-timber forest products of 2014)* (in Japanese). http://www.rinya.maff.go.jp/j/press/tokuyou/pdf/150929-01.pdf. Accessed 24 Feb 2017.

Chapter 5
Consumer Reaction and Willingness to Buy Food Produced Near the FDNPP

In this chapter, I would like to explain the consumer reaction toward buying rice, cucumbers, apples, raw shiitake mushrooms, livestock products (beef, pork, chicken, eggs), seafood products (tuna fish and wakame seaweed), and mineral water from regions near the Fukushima Daiichi Nuclear Power Plant (FDNPP). These agricultural products are chosen because they have relatively high production levels in Fukushima Prefecture. We investigate this consumer reaction by focusing on the following six factors that are likely to have an influence on the consumers' willingness to buy. The first factor is the consumers' eating habits, the second is their perceptions of food safety, the third is their interests in social problems, the fourth is their attitudes toward nuclear contamination, the fifth is their willingness to accept foods produced near the FDNPP, and the final factor is their social attributes.

The reason for investigating the willingness to buy for ten different food products is because the levels of radioactive contamination detected among different agricultural products were not the same after the Fukushima accident and that the consumer reaction might vary among various food products. Therefore, it would be interesting to find out the difference in the consumer reaction toward various food products.

In the next section, we will go over the details of the survey we performed to examine the consumer reactions.

In the second section, I will explain the factors affecting the consumer reaction toward cucumbers, apples, beef, and pork from regions near the FDNPP, which were food that a majority of the survey respondents had a positive willingness to buy.

The third section will discuss the factors affecting the consumers' willingness to buy shiitake mushrooms, chicken eggs, and tuna fish that about a half of the survey respondents were willing to accept them even if they were from regions near the FDNPP.

K. Aruga, *Consumer Reaction, Food Production and the Fukushima Disaster*,
DOI 10.1007/978-3-319-59849-9_5

In the fourth section, I will show the results of the factors that influence consumer reaction for rice, wakame seaweed, and mineral water, which were food that a majority of the respondents were not willing to buy if the products were from regions near the FDNPP.

In the final section of this chapter, I will summarize the factors that are likely affecting the consumer's willingness to buy food produced near the FDNPP based on the overall analysis of the survey.

5.1 Overview of the Consumer Survey

The consumer survey data used in this book was conducted between January 30 and February 4 of the year 2014. The title of the survey was "Survey on the Food Safety Issue Related to Radioactive Contamination," and it was performed on consumers from all parts of Japan. The survey was done with the help of a marketing survey company using an internet survey system. The respondents of the survey were selected based on the resident registration database of Japan, so the distribution of the respondents is similar to the actual population distribution of Japan.

5.1.1 Questions Related to Eating Habits

The subject of the survey is general consumers whose ages ranged between 20 and 69. The survey consists of the following six parts.

The first part is related to the consumers' eating habits such as their grocery shopping frequency, eating style, factors consumers are concerned about when buying food products, and so on (see Table 5.1). This part of the survey is constructed to understand how differences in the people's eating habits and food related lifestyle like shopping frequency will affect their willingness to buy food produced in regions near the FDNPP.

In each survey, we asked questions about two products that were chosen among the ten products (nine agricultural products and mineral water) of our interest. For example, when we wanted to ask the consumers about beef and pork, we displayed the name of these products in the "Good 1" and "Good 2" of Q2 and Q3 in Table 5.1. The combinations of two products the respondents answered are "rice and mineral water," "cucumbers and apples," "shiitake mushrooms and chicken eggs," "beef and pork," and "tuna fish and wakame seaweed." We randomly displayed one of these five different combinations of food products in the questionnaire, and the number of samples for each combination was equalized to minimize the dispersions among the combinations.

Table 5.1 Questions related to eating habits, food safety issues, and interests in social problems

Questions related to eating habits	
Q1	How often do you buy food products for yourself or for your family? 1. Almost every day 2. A few days a week 3. A few days a month 4. Not so often 5. Never
Q2	Have you ever had a chance to buy "Good 1"? 1. Yes 2. Never
Q3	Have you ever had a chance to buy "Good 2"? 1. Yes 2. Never
Q4	How often do you cook a meal for yourself or for your family? 1. Almost every day 2. A few days a week 3. A few days a month 4. Not so often 5. Never
Q5	How do you usually take your meal? 1. Do not use ready to eat food and eat cooked meal at home 2. Have ready to eat food 3. Meals prepared at convenience stores, department stores, grocery stores, and so on 4. Eat at a fast food restaurant 5. Eat at a normal restaurant 6. Other ways. Please write in detail: ____________
Q6	What do you concern the most when you buy a food products? Please pick the answer that suits you the most 1. Quality of the food 2. Freshness of the food 3. Expiration date and best buy date 4. Safety of food 5. Price 6. Other factors. Please write in detail: ____________
Questions related to food safety issues	
Q7	What do you usually do to keep yourself away from unsafe food? 1. Do not do anything special 2. Try to buy organic food 3. Check whether the food contains genetically modified organism 4. Try not to buy products that contain additives 5. Check the product origin 6. Check the expiration and best buy dates 7. Other answers. Please write in detail: ____________
Q8	How much would you trust the information you get from the labels on food? Please chose your answer based on a scale of 1 (do not trust at all) to 10 (very much trust the information)?
Questions related to interests in social problems	
Q9	How much are you willing to attend environmental activities such as preserving ecosystem, afforestation activities for mitigating climate change, and so on? Please chose your answer based on a scale of 1 (not willing at all) to 10 (very willing to attend)
Q10	How much are you willing to attend reconstruction activities that will help regions to recover from damages after the Tohoku-Pacific Ocean Earthquake? Please chose your answer based on a scale of 1 (not willing at all) to 10 (very willing to attend)

The reason for having questions for only two food products instead of including all ten products at once is because we repetitively asked a very similar question for different products so that we were concerned about the answers for the latter product to become biased. This kind of bias occurs because the respondents get tired of answering similar questions repeatedly and begin answering them halfheartedly when there are too many questions to answer with little distinction between them. Thus, if we include questions for many products in the survey, it is likely that questionnaire for products that are put in the latter part of the survey to become more biased. Such bias that arises due to differences in the position of questionnaires within a survey is called an order bias. This kind of order bias could happen even when asking the same question for two products. To avoid such bias, we created two sets of questionnaires by switching the orders of the two food products in our survey.

5.1.2 Questions Related to Food Safety Issues

The second part of the survey consists of questions related to food safety issues like what the consumers are concerned about when buying food products and how much the consumers trust the information they receive from the labels place on food products (see Table 5.1).

The purpose of having these questions in the survey is to understand the consumers' perceptions of food safety issue by clarifying their attitudes toward risky food products and the levels of trust they have on food labels. Then we will examine how people with a different view on the food safety issue have different reactions about buying food products from regions near the FDNPP.

5.1.3 Questions Related to Interests in Social Problems

In the third part of the survey, we asked the respondents to see their ideas about conserving the natural environment and supporting the regions that were hit by the tsunami and earthquake to recover from their damages (see Table 5.1).

Asking questions about how much the respondents consider social issues is helpful for understanding whether people's altruistic consciousness of helping the society will affect their reaction toward food products from regions near the FDNPP. I thought we might find some evidence that individuals who have more interests in social problems are more likely to buy products from regions near the FDNPP thinking their consumption behaviors will help these regions to recover the damages after the Tohoku-Pacific Ocean Earthquake.

5.1.4 Questions Related to Radioactive Contamination

As seen in Table 5.2, the fourth part of the survey contains questions about the respondents' perceptions of risk from radioactive food contamination, their knowledge about radiation, the origin of food products as to avoid the risk of

Table 5.2 Questions related to radioactive contamination

Q11	How do you perceive the risk of radioactive contamination of food sold at a grocery store? Please answer based on a scale of 1 (very low) to 10 (very high)?
Q12	Which of the following explanation about radiation do you know? Please choose all choices that apply 1. There are three types of radiation rays, alpha, beta, and gamma rays, and these rays have different abilities to pass through materials 2. Radioactive isotopes continue to decay until they become stable isotopes 3. Among the units to indicate the level of radioactive materials in food, sievert (Sv) represents the degree of influence on human body while becquerel (Bq) is used to measure the strength of radioactivity 4. Regardless of human activities, radiation exists in nature such as cosmic rays, earth rays, and so on 5. It is said that when additional amount of radiation received exceeds 100 mSv the probability of developing cancer in a lifetime increases about 0.5% 6. I do not know any of the above knowledge regarding radiation
Q13	Which of the following origin of product do you worry about the risk of radioactive contamination when buying food? Please choose all choices that apply 1. I do not care the origin of product when buying food 2. Fukuhisma Prefecture 3. Product origin of 100 km within the FDNPP 4. Product origin of 200 km within the FDNPP 5. Aomori Prefecture 6. Product origin of 300 km within the FDNPP 7. Product origin that is more than 300 km apart from the FDNPP
Q14	For person who chose other than 1 (I do not care the origin of product when buying a food product) in Q13, which of the food products do you care about the risk of radioactive contamination? Please pick three products from the following choices 1. Rice 2. Vegetable 3. Fruit 4. Eggs 5. Milk 6. Meat products such as beef, pork, and chicken 7. Fresh (raw) fish 8. Shellfish 9. Seaweed 10. Mushrooms 11. Tea 12. Processed food and confectioneries
Q15	Do you trust the current safety standards for radioactive cesium in food? Please answer based on a scale of 1 (do not trust at all) to 10 (very much trust the standard)

radioactive contamination, and finally, the respondents' trust toward the current national safety standards for radioactive cesium in food.

These questions are meant to grasp the relationship between the respondents' consciousness toward radioactive contamination and their willingness to buy food produced in regions near the FDNPP.

5.1.5 Questions Related to the Respondents' Willingness to Buy Food Produced Near the FDNPP

The fifth part of the survey asked the respondents about their willingness to buy food produced in regions near the FDNPP by questioning at what price difference with food from regions apart from the FDNPP they will accept to purchase food of these regions (see Table 5.3).

In this questionnaire, we compared the respondents' willingness to buy food of regions 100 km apart from the FDNPP and that of regions 300 km apart from the FDNPP. For the respondents to have ideas about the distance from the FDNPP, we showed them a figure like the one in Fig. 5.1 and the respondents were able to see this figure while answering the questions.[1]

The reason we used the direct distance from the FDNPP such as 100 km or 300 km apart from the FDNPP in the question to determine the respondents' willingness to buy food from regions near the FDNPP is because the restricted areas like the warning zone (see Table 2.2) was established using the linear distance from the FDNPP. Another reason for applying distance in our willingness to buy questionnaire is because it would be easier for the respondents who do not know the name of the prefectures or geography of regions near the FDNPP to have ideas about the distance of locations from the FDNPP.

Here too, like the Q2 and Q3 of Table 5.1, we first asked the Q16–Q18 questions for the first product and then asked the same questions for the second product through Q19–Q21 (see Table 5.3). The combinations of these two goods are the same as those explained in Sect. 5.1.1.

To clarify the differences among beef, pork, chicken eggs, shiitake mushrooms, tuna fish, and wakame seaweed produced in regions 100 and 300 km apart from the FDNPP, we added in the footnote some additional information about these foods. For beef, pork, and chicken eggs we added an assumption that the beef, pork, and chicken are fed with meals produced from the regions. For shiitake mushrooms we put a footnote saying that the shiitake mushrooms are produced with hardwoods grown in the regions. Finally, for tuna fish and wakame, we added that they are harvested from the closest sea of the region.

---

[1]In the actual survey performed on January 30 through February 4, 2014, we showed the Google map with names of cities and concentric circles of 100, 200, and 300 km around the FDNPP.

Table 5.3 Questions related to respondents' willingness to buy food produced near the FDNPP

Questions related to respondent's willingness to buy food produced near the FDNPP	
Q16	Please see Fig. 5.1, which shows the distance from the FDNPP. Imagine a situation where "Good 1" that is produced in a region 100 km away from the nuclear plant and the same product from a region 300 km away from the FDNPP are sold next to each other in a grocery store. When no radioactive materials above the current national food safety standards are detected after a radiation test is performed for "Good 1," are you willing to buy "Good 1" from a region 100 km away from the nuclear plant if "Good 1" was sold at a same price with the "Good 1" from a region 300 km away from the FDNPP? 1. Yes 2. No
Q17.1	If you answered that you do not want to buy "Good 1" in Q16, will you buy "Good 1" from a region 100 km away from the nuclear plant when the discount rate of "Good 1" was 10% compared with "Good 1" from a region 300 km away from the FDNPP? 1. Yes 2. No
Q17.2	If you answered that you do not want to buy "Good 1" in Q16, will you buy "Good 1" from a region 100 km away from the nuclear plant when the discount rate of "Good 1" was 20% compared with "Good 1" from a region 300 km away from the FDNPP? y1. Yes 2. No
Q17.3	If you answered that you do not want to buy "Good 1" in Q16, will you buy "Good 1" from a region 100 km away from the nuclear plant when the discount rate of "Good 1" was 30% compared with "Good 1" from a region 300 km away from the FDNPP? 1. Yes 2. No
Q17.4	If you answered that you do not want to buy "Good 1" in Q16, will you buy "Good 1" from a region 100 km away from the nuclear plant when the discount rate of "Good 1" was 40% compared with "Good 1" from a region 300 km away from the FDNPP? 1. Yes 2. No
Q17.5	If you answered that you do not want to buy "Good 1" in Q16, will you buy "Good 1" from a region 100 km away from the nuclear plant when the discount rate of "Good 1" was 50% compared with "Good 1" from a region 300 km away from the FDNPP? 1. Yes 2. No
Q17.6	If you answered that you do not want to buy "Good 1" in Q16, will you buy "Good 1" from a region 100 km away from the nuclear plant when the discount rate of "Good 1" was 60% compared with "Good 1" from a region 300 km away from the FDNPP? 1. Yes 2. No
Q18	If you answered that you do not want to buy "Good 1" in Q17.6 (discount rate of 60%), will you buy "Good 1" from a region 100 km away from the nuclear plant when there was a label on the product showing that the product has no risk of radioactive contamination? 1. Yes 2. No
Q19	The same question as Q16 for "Good 2" (the "Good 1" in Q16 is replaced to "Good 2")
Q20.1–Q20.6	The same question as Q17.1–Q17.6 for "Good 2" (the "Good 1" in Q. 17.1–Q17.6 is replaced to "Good 2")
Q21	The same question as Q18 for "Good 2" (the "Good 1" in Q18 is replaced to "Good 2")

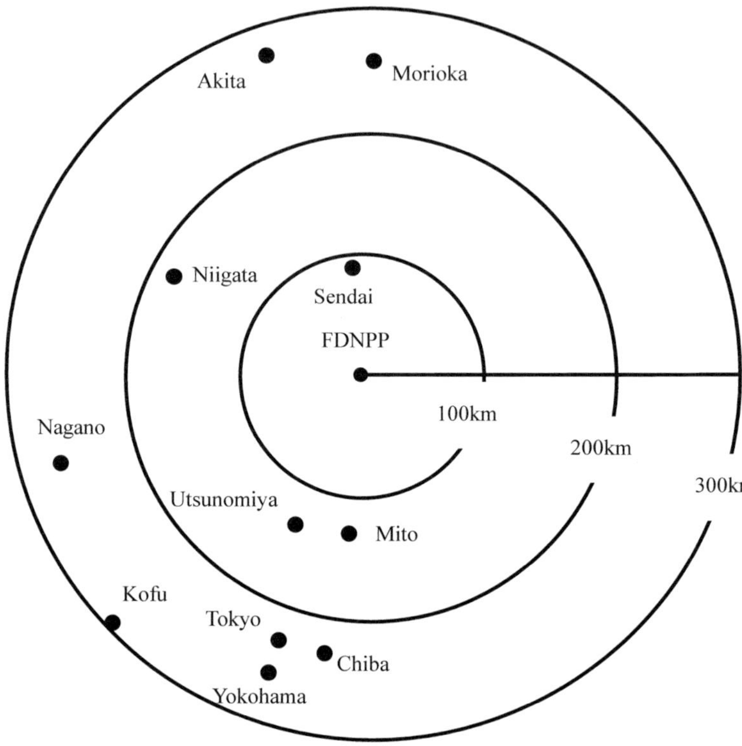

Fig. 5.1 Distance from the FDNPP

5.1.6 Questions Related to Social Attributes

In the last part of the survey, I asked the respondents' social attributes. As seen in Table 5.4, questions to learn the respondents' resident locations, gender, age, family members, education background, and annual income are included in this part.

In regards to the respondents' residence, we found from the survey results that a majority of the respondents lived in the three largest metropolitan areas of Japan: Tokyo, Nagoya, and Osaka areas. This is perhaps related to the fact that more than half of the whole population of Japan belong to these three metropolitan areas as of January 2014.

For questions related to the family members of the respondents, we included a questionnaire to ask the number of children within their families. We added this questionnaire because we expected that respondents with more children are likely more anxious about the risk of radioactive contamination. It is understood that the rate of thyroid cancer among children has increased after the Chernobyl accident.

Table 5.4 Questions related to social attributes

Q22	Please write the zip code of your residence
Q23	What is your gender? 1. Male 2. Female
Q24	What is your age? 1. 18–29 2. 30–39 3. 40–49 4. 50–59 5. 60–69
Q25	Please choose the family member you live with. Choose all choices that apply 1. An Infant 2. Elementary school age children 3. Junior high school age children 4. High school age children or children with similar age 5. Family member whose age is over 65 6. None of the above applies
Q26	How many of your family members are aged 15 or below?
Q27	What is your highest completed level of education? 1 = elementary and junior high school 2 = highschool 3 = junior college, 4 = some college credit, but no degree, 5 = bachelor's degree, 6 = master's degree, 7 = doctorate degree
Q28	What is your annual income of this year? 1 = less than 2 million yen 2 = 2–4 million yen 3 = 4–6 million yen 4 = 6–8 million yen 5 = 8–10 million yen 6 = 10–12 million yen 7 = 12–14 million yen 8 = over 14 million yen

Hence, we believe that respondents' characteristic of having more children probably has a considerable impact on their reaction toward food produced near the FDNPP.

5.1.7 Respondents' Gender and Age Distributions

As I explained in Sects. 5.1.1 and 5.1.5, the respondents of the survey answered the questionnaires related to food shopping experience and willingness to buy food from regions near the FDNPP based on five randomly selected combinations of goods. The combinations used in the survey are "rice and mineral water," "cucumbers and apples," "shiitake mushrooms and chicken eggs," "beef and pork," and "tuna fish and wakame seaweed." The total number of respondents for each of these food combinations can be seen in the last column of Table 5.5. As seen in the table, the number of respondents for "rice and mineral water" was 1710, "cucumbers and apples" was 1757, "shiitake mushrooms and chicken eggs" was 1710, "beef and pork" was 1768, and "tuna fish and wakame seaweed" was 1768. Hence, summing all five combinations, the total number of respondents in the survey was 8713.

The reason for having a different number of respondents among these five food combinations is because although the survey was conducted in a way to have the same number of respondents for each combination, the number of respondents who answered the survey was different among the combinations.

Table 5.5 Gender distributions of the respondents

	Male	Female	Total
Rice and mineral water	898 (53%)	812 (47%)	1710
Cucumbers and apples	877 (50%)	880 (50%)	1757
Raw shiitake mushrooms and chicken eggs	908 (53%)	802 (47%)	1710
Beef and pork	921 (52%)	847 (48%)	1768
Tuna fish and wakame seaweed	901 (51%)	867 (49%)	1768
Whole sample	4505 (52%)	4208 (48%)	8713

Table 5.6 Age distributions of the respondents

	20–29	30–39	40–49	50–59	60–69
Rice and mineral water	264 (15%)	387 (23%)	269 (16%)	312 (18%)	478 (28%)
Cucumbers and apples	279 (16%)	403 (23%)	266 (15%)	324 (18%)	485 (28%)
Raw shiitake mushrooms and chicken eggs	265 (15.5%)	372 (22%)	285 (16.5%)	313 (18%)	475 (28%)
Beef and pork	286 (16%)	383 (22%)	306 (17%)	291 (16.5%)	502 (28.5%)
Tuna fish and wakame seaweed	289 (16.5%)	388 (22%)	302 (17%)	340 (19%)	449 (25.5%)
Whole sample	1383 (16%)	1933 (22%)	1428 (16.5%)	1580 (18%)	2389 (27.5%)

Table 5.5 shows the gender distribution of the survey respondents among the food products investigated in this book. Although there were more male respondents in the survey, you can see from the table that the ratio between the genders was almost the same in all food products.

As the last part of this subsection let's look at the age distribution of the respondents. As seen in Table 5.6, about 15–16% of the respondents are in their 20s, about 22–23% are in their 30s, 15–17% are in their 40s, 16–19% are in their 50s, and finally 25–28% are in their 60s. The reason for having a higher number of percentage for respondents of age 60s is perhaps because the percentage of the elderly people are increasing in Japan. In fact, the number of people who are aged 65 or over was 31.86 million in 2013 and this accounts for 25% of Japan's population (MIC 2013).

5.1.8 Food that the Respondents Consider the Product Origin

From the next section, we try to find out the factors that affect consumers' willingness to buy for ten food products. For this purpose, we classify the ten food

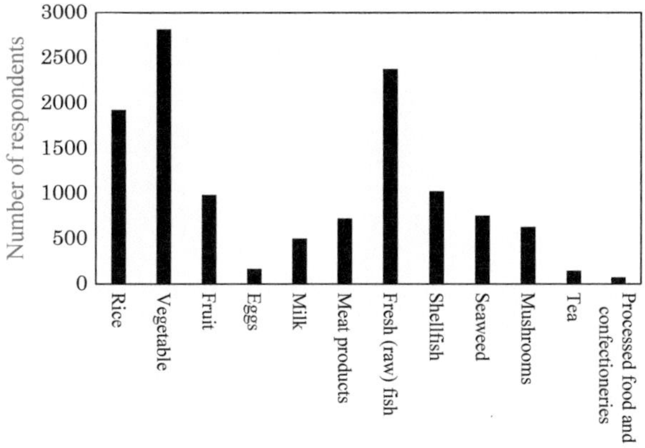

Fig. 5.2 Food that respondents worry about its product origin

products of our interest into three categories. The first category is food from regions near the FDNPP that a majority of respondents choose to buy. The second category is food about a half of the respondents agree to buy. The third one is food that a majority of respondents are not willing to buy if the food was from regions near the FDNPP. Before going to this section, I would like to discuss if there is any difference in the consumer reaction among the types of food.

Figure 5.2 shows the results of food that consumers feel anxious about its product origin for different types of food. The figure is created based on the answers of respondents who replied that they have food that they would avoid to buy if their product origin is from certain areas. The figure shows the number of respondents for each food in the choices provided in Q14 of Table 5.2. The food with more respondents in the figure are the ones that the respondents have concerns about its product origin. The figure suggests that more people worry about its product origin for rice, vegetable, and fresh fish while people do not care much about its product origin for eggs, tea, and processed food and confectioneries. The figure indicates that there are differences in the type of food that consumers feel anxious about its product origin. This is why we classified the ten food products into three categories by the differences in the degree of willingness to buy food from regions near the FDNPP.

5.1.9 Classification of the Ten Products Based on the Respondents' Willingness to Buy

Here, I would like to explain the criteria we used to classify the ten products into three categories. The first category is food from regions near the FDNPP that a majority of respondents choose to buy. The second category is food from regions

Table 5.7 Differences in the respondents' WTA among different food products

Rice, cucumbers, apples, and shiitake mushrooms				
	Rice	Cucumbers	Apples	Shiitake mushrooms
Buy	48.8%	55.0%	54.9%	49.6%
Not buy	51.2%	45.0%	45.1%	50.4%
Number of samples	1710	1757	1757	1710

Livestock products			
	Beef	Pork	Chicken eggs
Buy	53.2%	52.3%	50.5%
Not buy	46.8%	47.7%	49.5%
Number of samples	1768	1768	1710

Seafood and mineral water			
	Tuna fish	Wakame seaweed	Mineral water
Buy	50.5%	46.2%	46.7%
Not buy	49.5%	53.8%	53.3%
Number of samples	1768	1768	1710

near the FDNPP that about half of respondents agree to buy, and the third one is food that a majority of respondents avoid to buy.

Table 5.7 depicts the results of the questions on whether the respondents agree to buy food from regions near the FDNPP at the same price with that from regions apart from the FDNPP. The table is created by summarizing the results of the questions Q16 and Q19 of Table 5.3.

According to the table, a majority of the respondents agreed to buy cucumbers, apples, beef, and pork from regions near the FDNPP at the same price with those produced in regions apart from the FDNPP. Thus, it is believable that for these food products there is no avoiding behavior among the respondents even if the foods were produced in regions close to the FDNPP and most of the respondents choose to buy the foods.

Next, for raw shiitake mushrooms, chicken eggs, and tuna fish from regions near the FDNPP, about a half of the respondents are willing to buy them at the same price with those from regions apart from the FDNPP (see Table 5.7). Hence, we categorize these food products as food that about a half of respondents are willing to buy.

Finally, the last food category belongs to food that a majority of respondents avoid to buy. As seen in Table 5.7, more than half of the respondents answered not willing to purchase products such as rice, wakame seaweed, and mineral water from regions near the FDNPP. Thus, from now on, we will classify these products as food that a majority of respondents avoid to buy.

5.2 Food Produced in Regions Near the FDNPP that a Majority of the Respondents Are Willing to Buy: Cucumbers, Apples, Beef, and Pork

As I explained in the previous chapter, the survey focused on the following six factors: respondents' eating habits, perceptions of food safety, interests in social problems, attitudes toward nuclear contamination, consumers' willingness to accept food produced near the FDNPP, and social attributes. In this section, we will examine how these six factors affect people's willingness to buy cucumbers, apples, beef, and pork. These four agricultural products are the ones that the majority of respondents chose to buy even if they are produced near the FDNPP.

I will compare the answers of respondents who answered yes to the question, "do you want to buy food produced 100 km away from the FDNPP if the price is the same as that produced 300 km away?" and those of respondents who selected "no, I do not want to buy it if it is the same price" (these questions are based on Q16–Q19 of Table 5.3). Finally, I will investigate how the above mentioned six factors are affecting the respondents' willingness to buy food produced near the FDNPP.

5.2.1 Respondents' Eating Habits and Their Willingness to Buy

Here, I would like to discuss the relationship between the respondents' eating habits and their willingness to buy cucumbers, apples, beef, and pork from regions near the FDNPP by showing the results of Fig. 5.3 through Fig. 5.7.

Shopping frequency and willingness to buy

Figure 5.3 shows the relationship between how often people buy grocery for themselves and their family members and their willingness to buy cucumbers, apples, beef, and pork produced near the FDNPP. The figure is created based on the question related to shopping frequency (Q1 of Table 5.1) and the respondents' answers for willingness to buy food produced near the FDNPP at the same price as that produced away from FDNPP (Q16 and Q19 of Fig. 5.3).

Figure 5.3 indicates that in all categories of shopping frequency (almost every day, several days a week, and several days a month), more people are willing to buy cucumbers and apples produced near the FDNPP even if they are sold at the same price as those produced apart from the FDNPP. This implies that shopping frequency does not affect people's willingness to buy cucumbers and apples from regions near the FDNPP.

On the other hand, for beef and pork, people who go shopping every day are not willing to buy these products if they are from regions near the FDNPP. We find that people who often go shopping for food are more concerned about the risk of

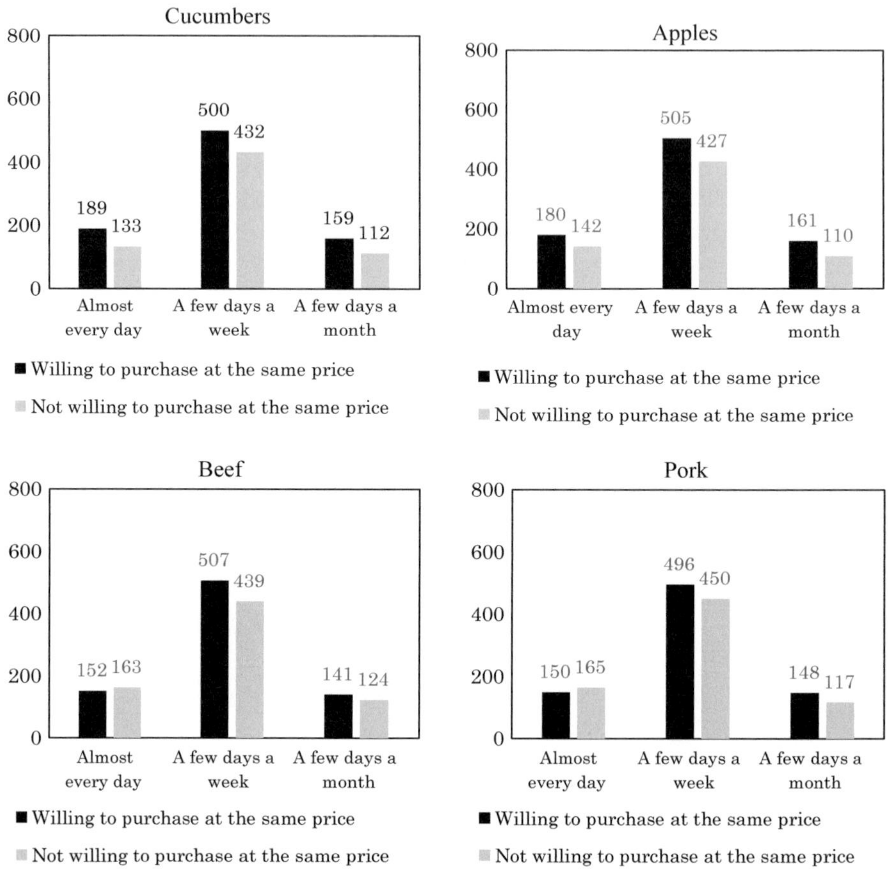

Fig. 5.3 Shopping frequency and the willingness to buy I

radioactive contamination. Therefore, it is presumable that people who frequently buy food tend to avoid beef and pork produced near the FDNPP compared with cucumbers and apples of those regions.

The reason more people avoid buying beef and pork compared with cucumbers and apples might be related to the fact that test samples of cucumbers and apples have never contained radioactive cesium above the level of national safety standards. High levels of radioactive cesium have been detected in some of the test samples for beef and pork.

We find that those who often go shopping to buy grocery are more likely to avoid buying food produced near the FDNPP compared with those who do not go shopping so frequently. This can be explained by the fact that frequent shoppers have more risk of buying food contaminated with radioactive material than non-frequent shoppers, and hence, they are very sensitive to the risk of consuming food contaminated with radioactive material.

Purchase experience and the willingness to buy

Next, we will discuss the differences in the willingness to buy cucumbers, apples, beef, and pork from regions near the FDNPP between the respondents who have experiences of buying the food for themselves and their family and those who never buy them. We will use Fig. 5.4 for this purpose based on questions related to shopping experience (Q2 and Q3 of Table 5.1) and respondents' willingness to buy food from regions near the FDNPP at the same price with that from regions apart from the FDNPP (Q16 and Q19 of Table 5.3).

According to Fig. 5.4, in every product there is not much difference in the respondents' willingness to buy food from regions near the FDNPP between the respondents who purchase the food by themselves and those who do not purchase the food. All the figures of cucumbers, apples, beef, and pork show that in both respondents with experience of buying the product and those without such experience, more respondents are willing to buy food produced in regions near the

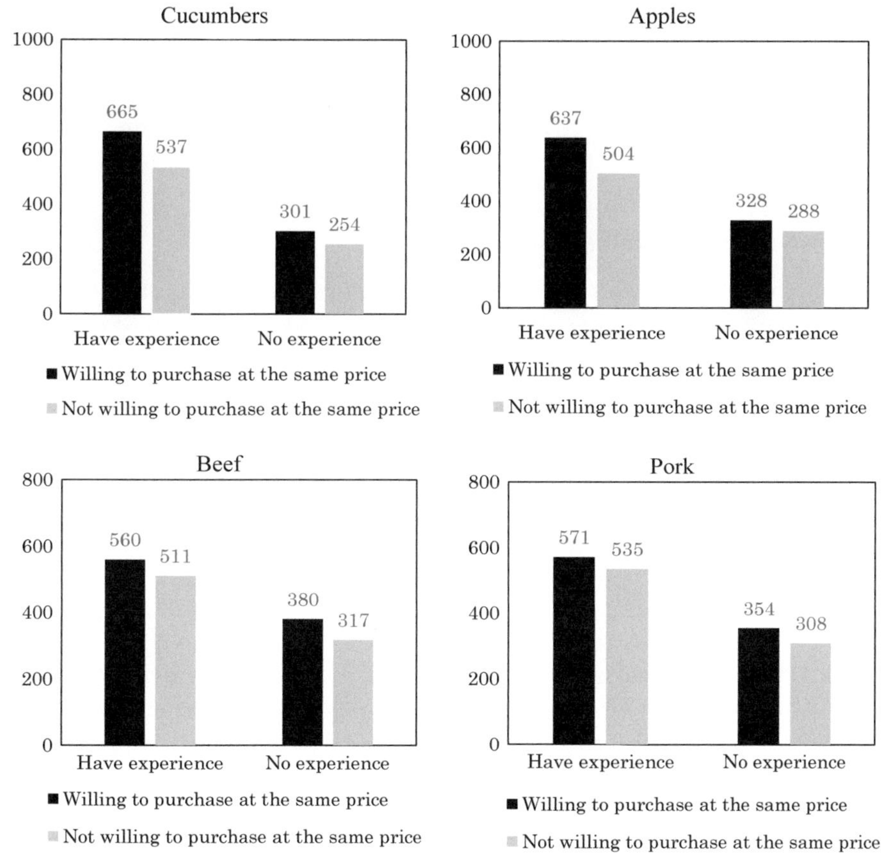

Fig. 5.4 Purchase experience and the willingness to buy I

FDNPP at the same price as food from regions apart from the FDNPP. This indicates that for foods that a majority of respondents reacted positively about buying them, consumers' purchase experiences do not affect their willingness to buy food.

Cooking frequency and the willingness to buy

Figure 5.5 presents the relationship between respondents' cooking frequency and their willingness to buy cucumbers, apples, beef, and pork from regions near the FDNPP. The figure is created from Q4 of Table 5.1 and Q16 and Q19 of Table 5.3.

The figure suggests that for all products the frequency of cooking does not seem to affect the respondents' willingness to buy food from regions near the FDNPP because in nearly all cooking frequency, from cook almost every day to never cook categories, more people are willing to buy food from regions near the FDNPP.

However, it is noticeable in the figure for pork that more people are not willing to buy food from regions near the FDNPP when they belong to the category cook almost every day. You can also see from the figure for other foods that the difference in the number of respondents who responded "yes" and "no" regarding willingness to buy is smaller among respondents who cook frequently compared with those who do not cook so frequently. Hence, this implies that for respondents who cook very often, meaning that their cooking frequency is high, the number of consumers who are willing to buy food from regions near the FDNPP becomes smaller compared with consumers that are not willing to buy food from these regions.

Eating style and the willingness to buy

Figure 5.6 illustrates the relationship between the respondents' eating style and their responses on willingness to buy food from regions near the FDNPP.

The "Mostly eat at home" category in the figure is based on the first choice of Q5 of Table 5.1, "Do not use ready to eat food and eat cooked meal at home." On the other hand, the "mostly eat out" category consists of meals that are not prepared at home like ready to eat meals, such as frozen meal, fast food, and so forth. This category is created using second to fifth choices from Q5 in Table 5.1. The figure compares the number of respondents who are willing to buy food from regions near the FDNPP and that of respondents who are not willing to buy such food and see if there is any difference between people who often cook meals at home and those who mostly go out for their meals.

From the figure, it is apparent that the difference in the eating styles of the respondents does not have much impact on the respondents' willingness to buy answers for cucumbers and apples. In both cucumbers and apples, regardless of their eating style, more people are willing to accept food from regions near the FDNPP. By contrast, for beef and pork, the number of respondents who avoid food from regions near the FDNPP becomes larger among those that mostly go out for their meals. The reason for respondents who often eat out to respond negatively toward food produced near the FDNPP is perhaps because recently the percentage

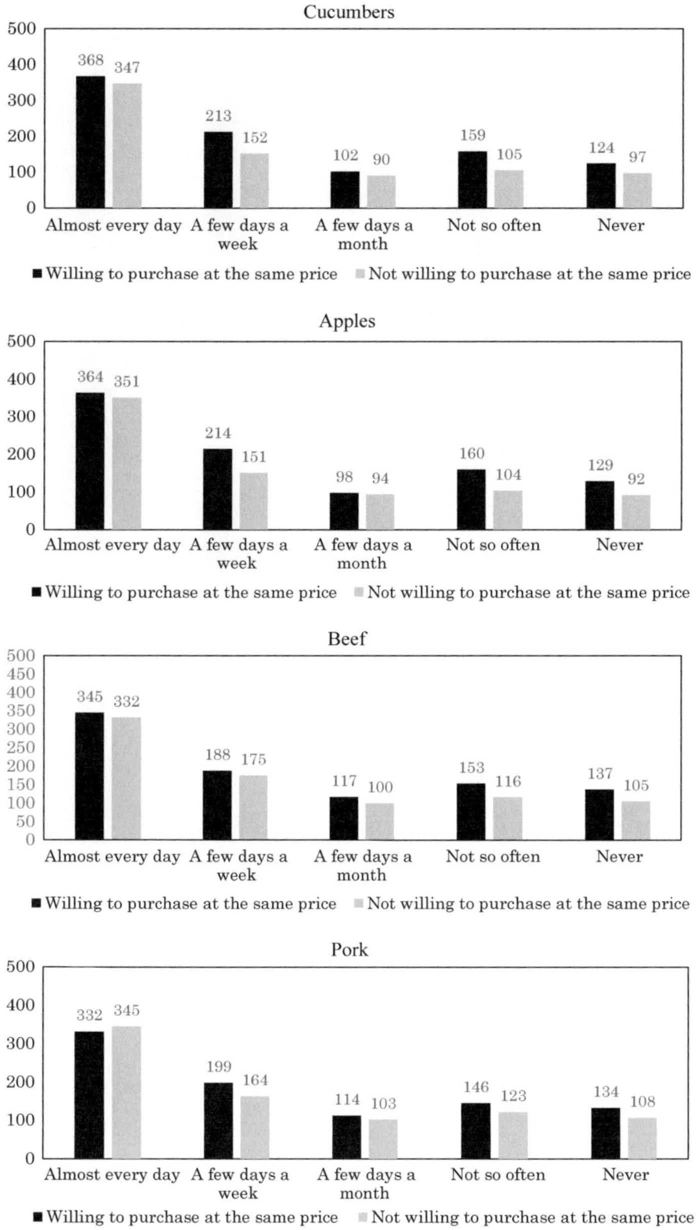

Fig. 5.5 Cooking frequency and the willingness to buy I

of restaurants using imported beef and pork is increasing and these people are not used to eating domestic products and that they are more sensitive to the risk of having food contaminated with radioactive material.

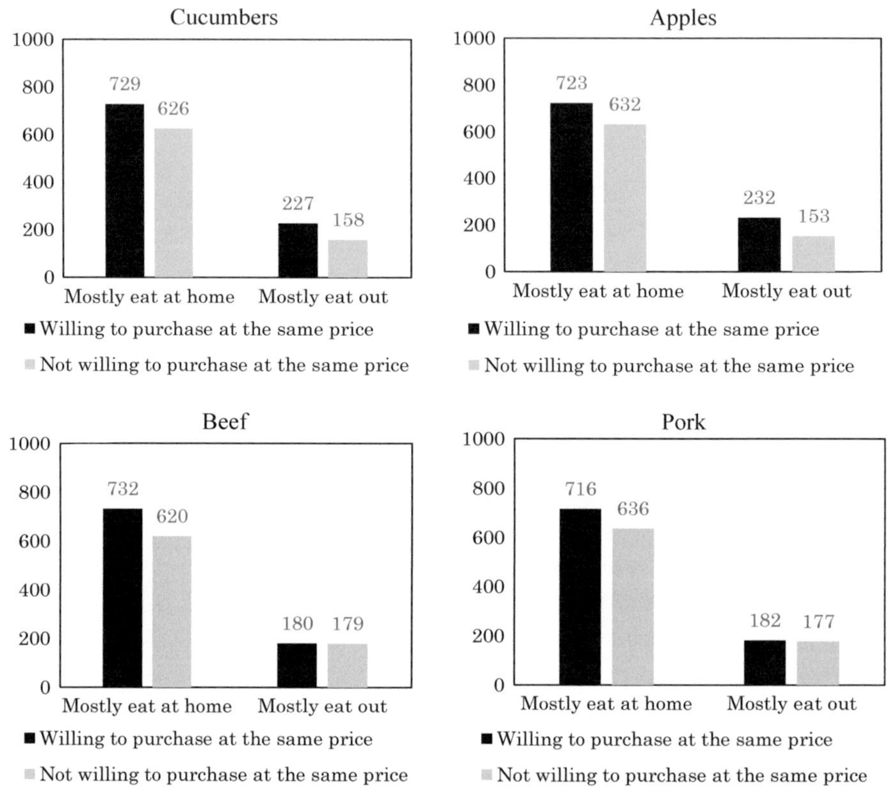

Fig. 5.6 Eating style and the willingness to buy I

Respondents' concerns when buying food and the willingness to buy

Figure 5.7 shows the relationship between the factors that consumers worry the most when purchasing food products and the consumers' response toward willingness to buy cucumbers, apples, beef, and pork from regions near the FDNPP. The component of factors that consumers are anxious about when buying food consists of food quality, freshness, expiration and best buy dates, food safety, price, and other factors as shown in Q6 of Table 5.1. Most of the respondents picked freshness and price among these factors.

Concerning the willingness to buy food from regions near the FDNPP, the number of respondents who answered yes to the willingness to buy question was higher in most of the factors in the figure except for safety of food. However, among the respondents who selected safety of food as factor that they put importance on when buying food, the percentage of people who are not willing to accept food from regions near the FDNPP was higher for apples, beef, and pork. This indicates that people who care about food safety issue are reluctant to buy food produced in regions near the FDNPP.

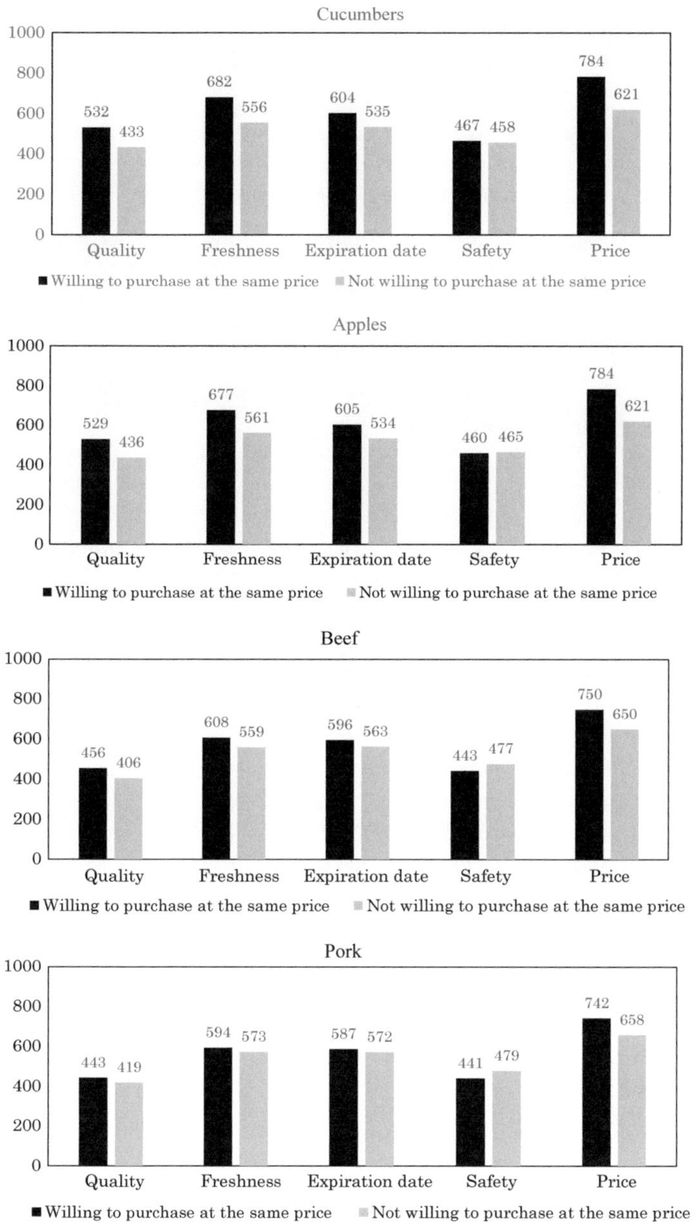

Fig. 5.7 Factors that consumers worry about when buying food and their willingness to buy I

5.2.2 Respondents' Perceptions of Food Safety Issue and Their Willingness to Buy

Second, in this subsection we will investigate the relationship between the consumers' perceptions of food safety and their willingness to buy cucumbers, apples, beef, and pork that are produced in regions near the FDNPP.

Respondents' perceptions of unsafe food and the willingness to buy

Figure 5.8 is a figure showing the relationships between what consumers normally do to keep themselves away from unsafe food and how categories in the figure affect their willingness to buy food from regions near the FDNPP for cucumbers, apples, beef, and pork. The categories related to consumers' actions for food safety are based on Q7 of Table 5.1.

According to the figure, people who answered "Try not to buy products that contain additives," "Check the product origin," and "Check the expiration and best buy dates" have positive reaction about buying cucumbers and apples produced in regions near the FDNPP at the same price as those from regions apart from the FDNPP.

On the other hand, among the people who picked "Try to buy organic food," and "Check whether the food contains genetically modified organism" among the choices in Q7 of Table 5.1, there were many respondents who answered negatively about buying cucumbers and apples from regions near the FDNPP.

For beef and pork, people who picked the category "Check the expiration and best buy dates" responded positively about buying those produced in regions near the FDNPP but in most other categories people responded negatively about buying beef and pork from regions near the FDNPP. It is likely that most individuals who do some kind of actions to keep themselves away from unsafe food are concerned about buying beef and pork from regions near the FDNPP.

Furthermore, in all food products of Fig. 5.8 you can see that people who do nothing special to keep themselves away from unsafe food have positive reaction about buying food from regions near the FDNPP.

In summary, the figure indicates that people who are more concerned about keeping themselves free from unsafe food are less willing to buy food from regions near the FDNPP.

Trust in food labels and the willingness to buy

Figure 5.9 illustrates the relationship between how much the respondents' trust the information they receive from the labels placed on food and the willingness to buy cucumbers, apples, beef, and pork from regions near the FDNPP. The figure is created by using the answers from Q8 in Table 4.1. The "Trust food label" category in the figure shows the number of respondents who selected 7 through 10 (very much trust the information) to the question, "How much would you trust the information you get from the labels on food?" The "Don't trust food label" category is the number of respondents who selected 1 (do not trust at all) through 4 to this question.

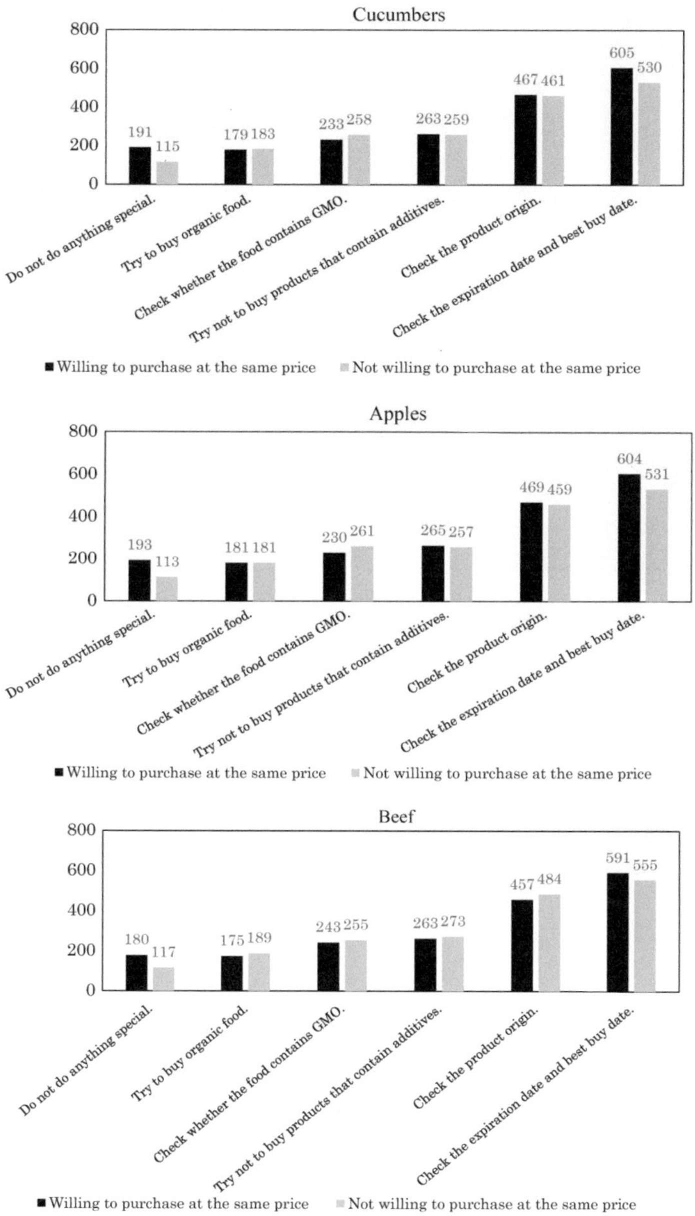

Fig. 5.8 Consumers' actions to keep themselves away from unsafe food and the willingness to buy I

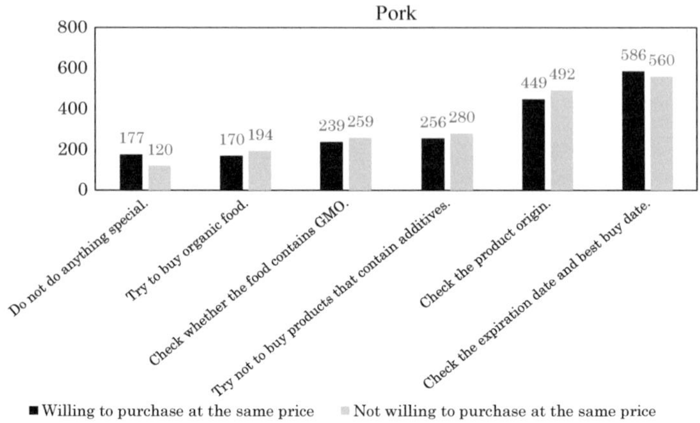

Fig. 5.8 (continued)

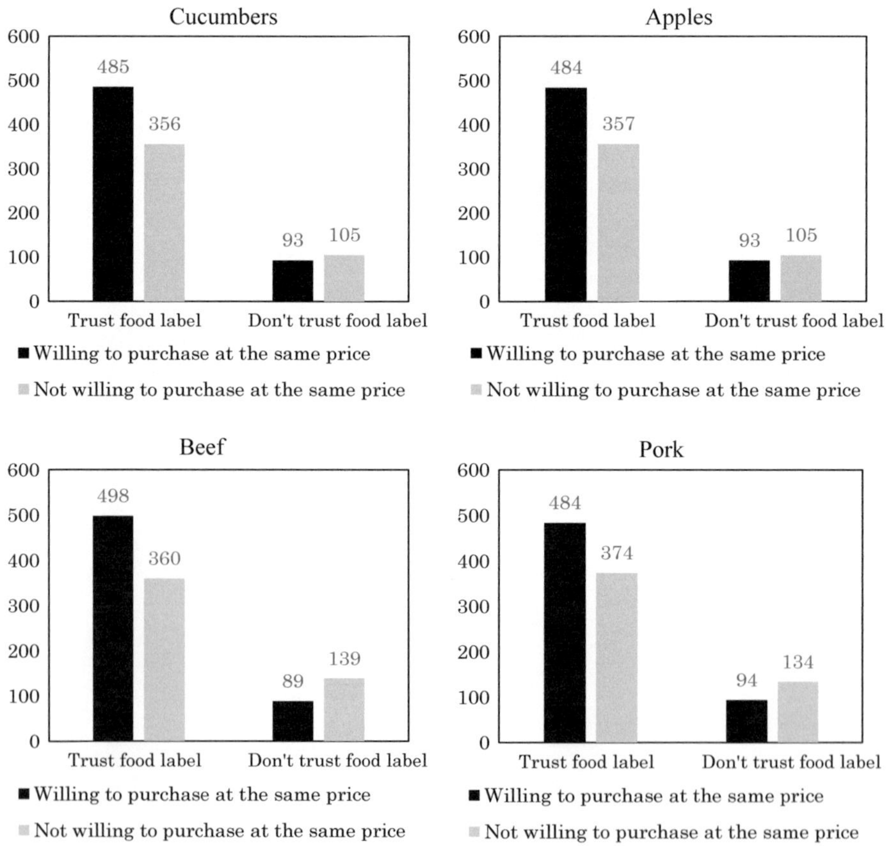

Fig. 5.9 Trust in food labels and the willingness to buy I

The figures suggest that more people have trust in food labels and respondents who trust the food labels are more willing to buy food from regions near the FDNPP. However, respondents who answered that they do not trust the food labels have negative reaction toward cucumbers, apples, beef, and pork produced in regions near the FDNPP. Hence, the results of the figure point out the importance of improving the reliability of the information provided on the food labels.

5.2.3 *Respondents' Interests in Social Problems and the Willingness to Buy*

Third, in this subsection we examine whether the respondents' degree of interests in social problems has effects on the respondents' willingness to buy cucumbers, apples, beef, and pork produced in regions near the FDNPP through Figs. 5.10 and 5.11.

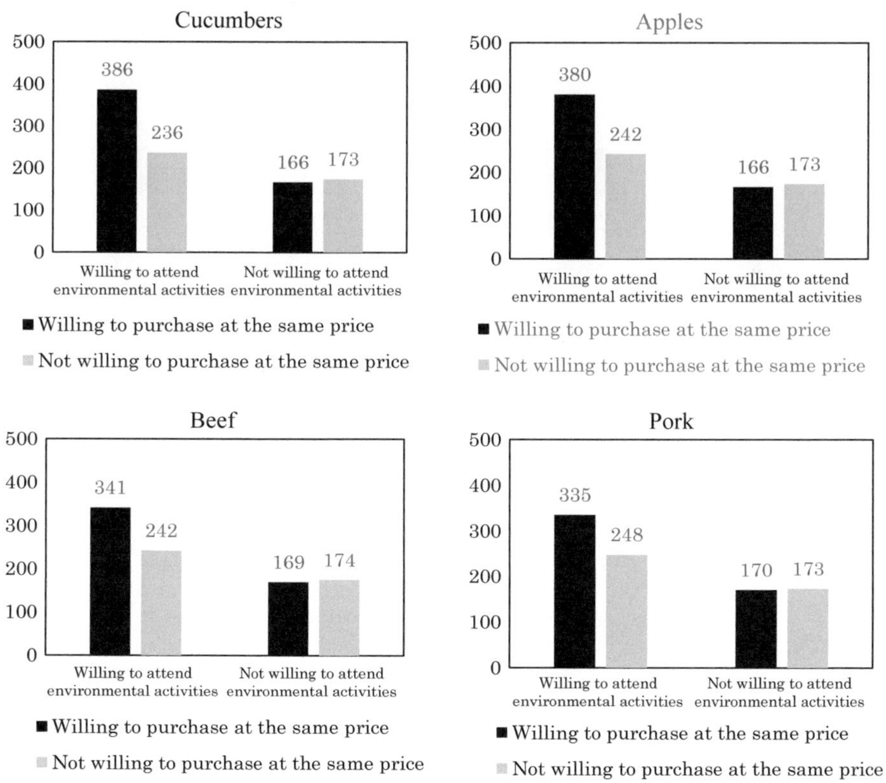

Fig. 5.10 Aggressiveness in preserving the environment and the willingness to buy I

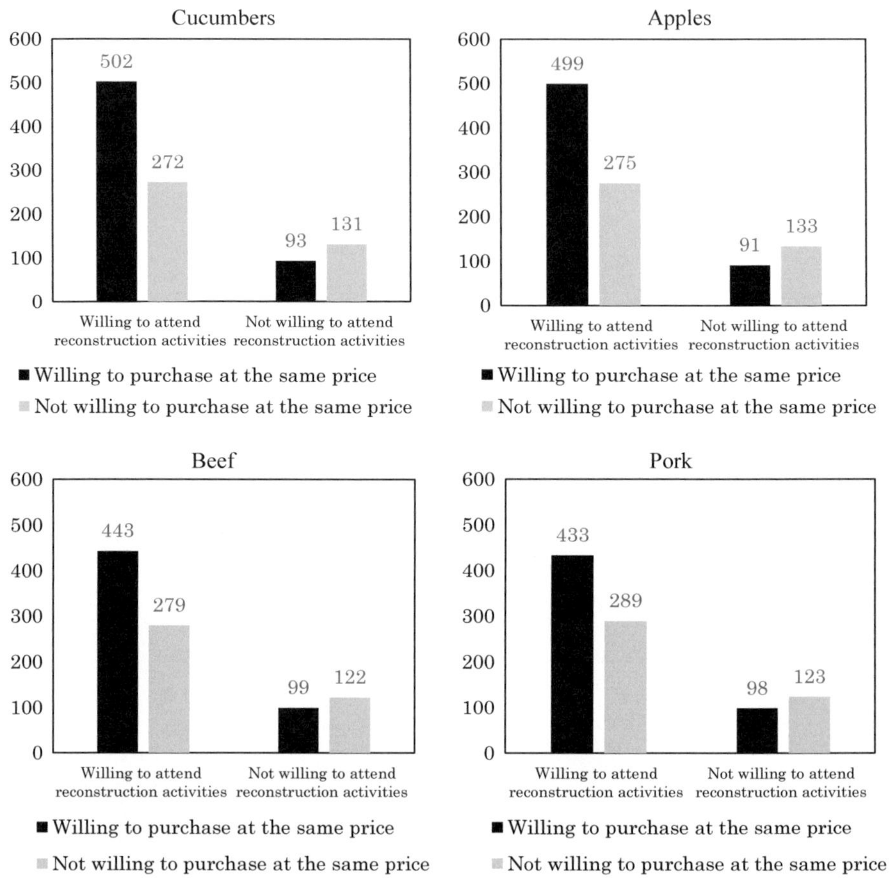

Fig. 5.11 Aggressiveness in reconstructing the regions suffered from the Tohoku-Pacific Ocean Earthquake and the willingness to buy I

Respondents' perceptions of environmental conservation and their willingness to buy

Figure 5.10 depicts the relationship between the respondents' perceptions about environmental conservation and their willingness to buy food from regions near the FDNPP. The respondents' perceptions about environmental conservation is formed by using the answers of Q9 of Table 5.1, "are you willing to attend environmental activities such as preserving ecosystem, afforestation activities for mitigating climate change, and so on?"

The category "willing to attend environmental activities" is the number of respondents who selected numbers 7 through 10 (very willing to attend) in Q9 of Table 5.1. Respondents that belong to this category are individuals who are serious about conserving the environment. On the contrary, the category "not willing to attend environmental activities" is the number of respondents who picked 1 (not willing at all) through 4, which represents people who are not so interested in solving environmental problems.

According to Fig. 5.10, consumers who are eager to assist in resolving environmental problems are more willing to buy cucumbers, apples, beef, and pork from regions near the FDNPP. By contrast, for all food products, the figure tells us that people who are half-hearted about joining activities to conserve the environment are not willing to buy food products from regions near the FDNPP at the same price as those from regions apart from the FDNPP. This result implies that consumers who have interests in social problems like environmental conservation are more willingness to buy foods produced in regions near the FDNPP.

Respondents' perceptions of helping the disaster regions and their willingness to buy

Next, we investigate the relationship between consumers' aggressiveness to help the regions that suffered from the Tohoku-Pacific Ocean Earthquake to recover from their damages and their willingness to buy cucumbers, apples, beef, and pork from regions near the FDNPP. The "willing to attend reconstruction activities" in Fig. 5.11 shows the number of respondents who choose numbers 7 through 10 (very willing to attend) in Q10 of Table 5.1, and the chart "not willing to attend reconstruction activities" describes the number of respondents who selected 1 (not willing at all) through 4 in Q10 of Table 5.1.

From Fig. 5.11, you can see that respondents who are anxious for attending reconstruction activities to help the disaster regions from the Tohoku-Pacific Ocean Earthquake are more willing to buy cucumbers, apples, beef, and pork produced in regions near the FDNPP compared with those who are passive about joining reconstruction activities. Although more people are enthusiastic about helping the disaster areas to recover from the earthquake, some people are uninterested in contributing to such areas, and among such people, it seems that more people have a tendency to avoid food that is from regions near the FDNPP.

Altruistic consumers and their willingness to buy

The results of Figs. 5.10 and 5.11 indicate that consumers who are more interested in contributing to solve social problems like environmental conservation and reconstruction activities of disaster areas are more likely willing to buy cucumbers, apples, beef, and pork produced in regions near the FDNPP.

It can be presumed that such consumers with interests in social problems having a higher willingness to buy food from regions near the FDNPP are related to altruism. As suggested by Aruga (2016), it is believable that such altruistic consumers are willing to buy food produced in regions near the FDNPP because they think buying more products from these regions will help these regions to recover from the disaster. Individuals whose level of satisfaction increase when other people's utility grow are known as altruistic people in environmental economics (Kolstad 2000) and it is likely that the results seen in Figs. 5.10 and 5.11 are related to altruistic people's behavior to help the disaster regions to recover from their suffering through purchasing products from these regions.

5.2.4 *Respondents' Perceptions of Radioactive Contamination and Their Willingness to Buy*

Fourth, in this subsection we will look into the relationship between the consumers' risk perception of food being contaminated with radioactive material and their willingness to buy from regions near the FDNPP using Fig. 5.12 through Fig. 5.15.

Respondents' risk perceptions toward radioactively contaminated food and their willingness to buy

Figure 5.12 shows how respondents of the survey think about the risk of food sold at grocery stores containing radioactive materials and how the differences in such risk perceptions will affect their willingness to buy cucumbers, apples, beef, and pork produced in regions near the FDNPP. The chart is created by defining the respondents who selected choices 7 through 10 (very high) from Q11 in Table 5.2 as individuals who think the risk of food sold at grocery stores to be contaminated with radioactive material is high. By contrast, the participants that picked 1 (very low) through 4 in Q11 is categorized as individuals who perceived the risk of radioactive contamination of food is low.

According to Fig. 5.12, consumers who think the risk of food sold at grocery stores to be contaminated with radioactive material is high are more likely to avoid buying food from regions near the FDNPP and such trend can be seen in all the figures: cucumbers through pork. On the other hand, among the consumers who perceive the risk of radioactive contamination of food sold at grocery stores is low, more consumers are willing to buy food from regions near the FDNPP, showing a positive reaction toward food from these regions.

Hence, the results of the figure imply that whether the consumers are willing to buy cucumbers, apples, beef, and pork from regions near the FDNPP depends on the way the consumers observe the risk of radioactive contamination of food they buy at grocery stores.

Respondents' knowledge of radiation and their willingness to buy

Figure 5.13 illustrates how consumers' knowledge related to radiation and radioactive material will influence the respondents' willingness to buy cucumbers, apples, beef, and pork produced in regions near the FDNPP. The categories in the horizontal axis are the choices for the radiation knowledge the respondents picked in Q12 of Table 5.2 and the chart shows the total number of respondents chosen for each category.

From the figure for cucumbers and apples, you can see that people who picked at least one of the radiation knowledge categories and had some knowledge of

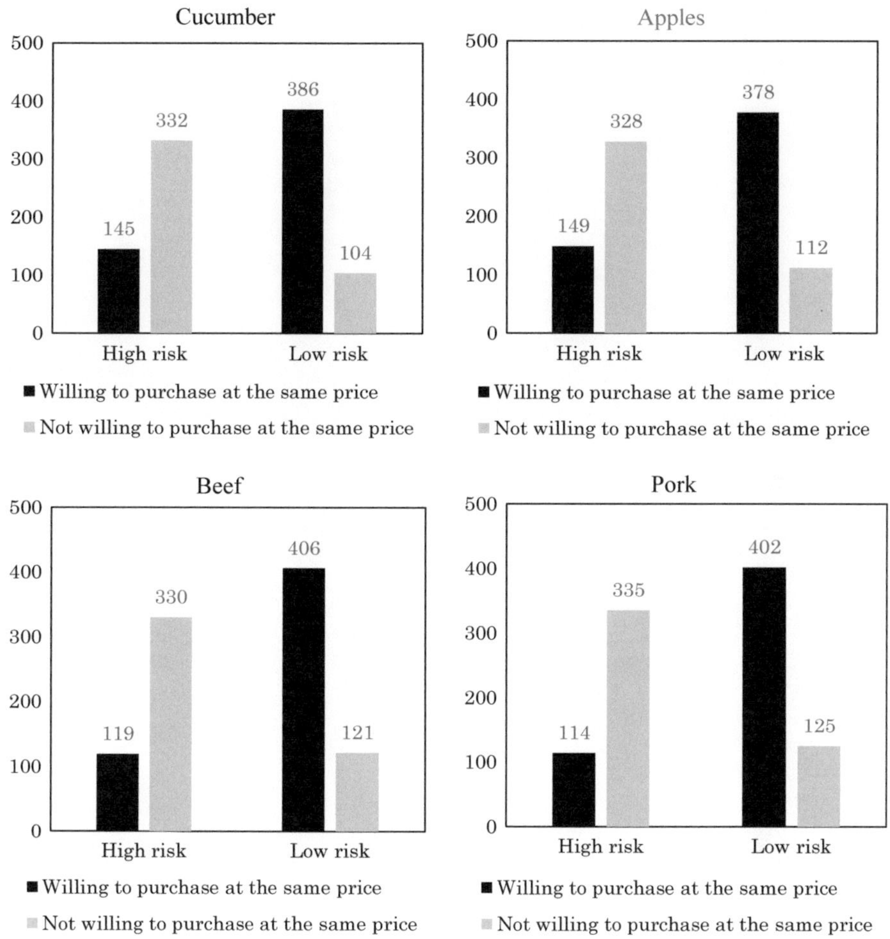

Fig. 5.12 Risk perceptions of food being contaminated with radioactive materials and the willingness to buy I

radiation or radioactive material are more willing to accept buying food from regions near the FDNPP. However, for beef and pork the number of people reacted negatively toward foods from regions near the FDNPP outnumbered those who reacted positively about buying such foods among the individuals who knew that "when an additional amount of radiation received exceeds 100 mSv the probability of developing cancer in a lifetime increases about 0.5%." Furthermore, people who answered that they know nothing about the radiation knowledge asked in the survey are also more likely to respond negatively about buying food from regions near the FDNPP.

It can be implied from the results of Fig. 5.13 that people who have little knowledge regarding radiation and radioactive material have tendencies to perceive

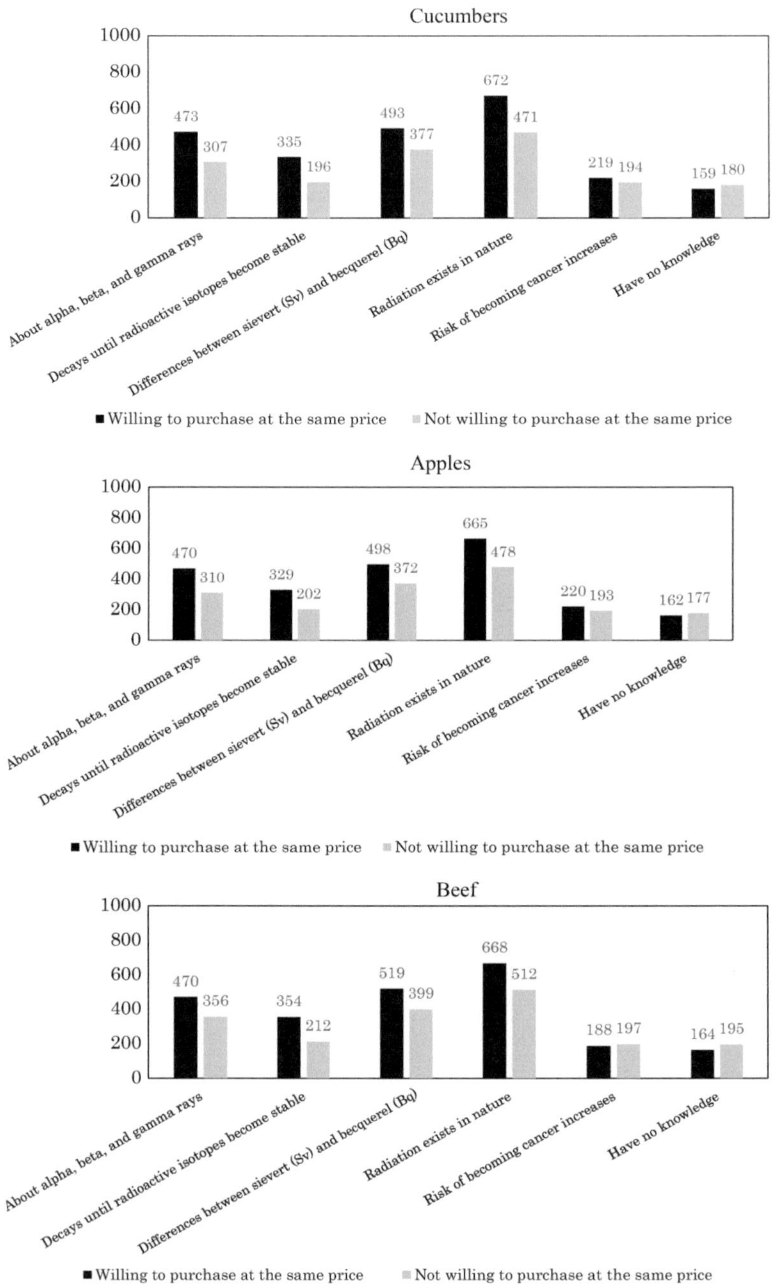

Fig. 5.13 Knowledge of radiation and the willingness to buy I

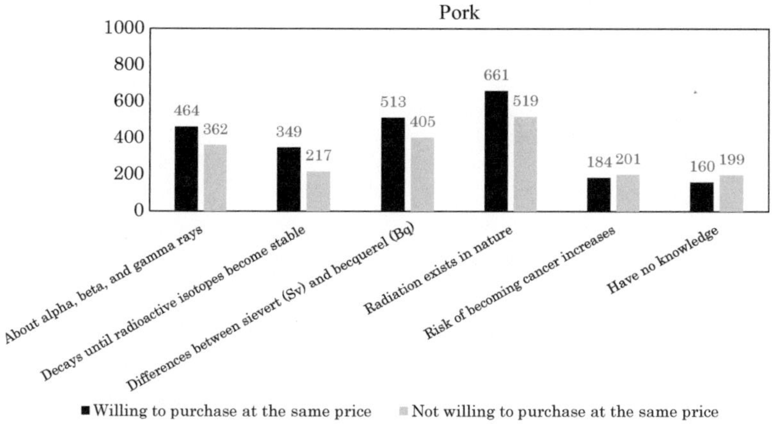

Fig. 5.13 (continued)

highly of the risk of food produced in regions near the FDNPP compared with people who have some knowledge about radiation. Hence, conducting policies like promoting education programs for radiation to have more people learn about radiation and radioactive material would be helpful to improve the risk perception toward food produced in regions near the FDNPP.

Respondents' perceptions of product origins and the willingness to buy

Next, we will go over the product origin that the respondents are concerned about the risk of radioactive contamination when buying food products and how their perceptions of product origins have effects on respondents' willingness to buy food from regions near the FDNPP. Figure 5.14 is built based on the respondents' answers for Q13 of Table 5.2, which asked the respondents about product origins that they would worry the risk of radioactive contamination when buying food. The categories shown in the horizontal axis are the choices provided in Q13.

You can see from Fig. 5.14 that respondents who answered they are worried about the risk of radioactive contamination of food from certain regions are mostly not willing to buy food from regions near the FDNPP. The figure also shows that among the categories of product origins presented in the figure, people worry the most when the food is from Fukushima Prefecture.

By contrast, there are quite a few individuals who answered they have no product origin that they are worried about when buying food products and a majority of these people have positive reaction about buying cucumbers, apples, beef, and pork from regions near the FDNPP.

Thus, people who believe that food from certain areas has a risk of radioactive contamination are more likely to avoid buying food from regions near the FDNPP but those who are indifferent about product origins have positive reaction toward food from regions near the FDNPP.

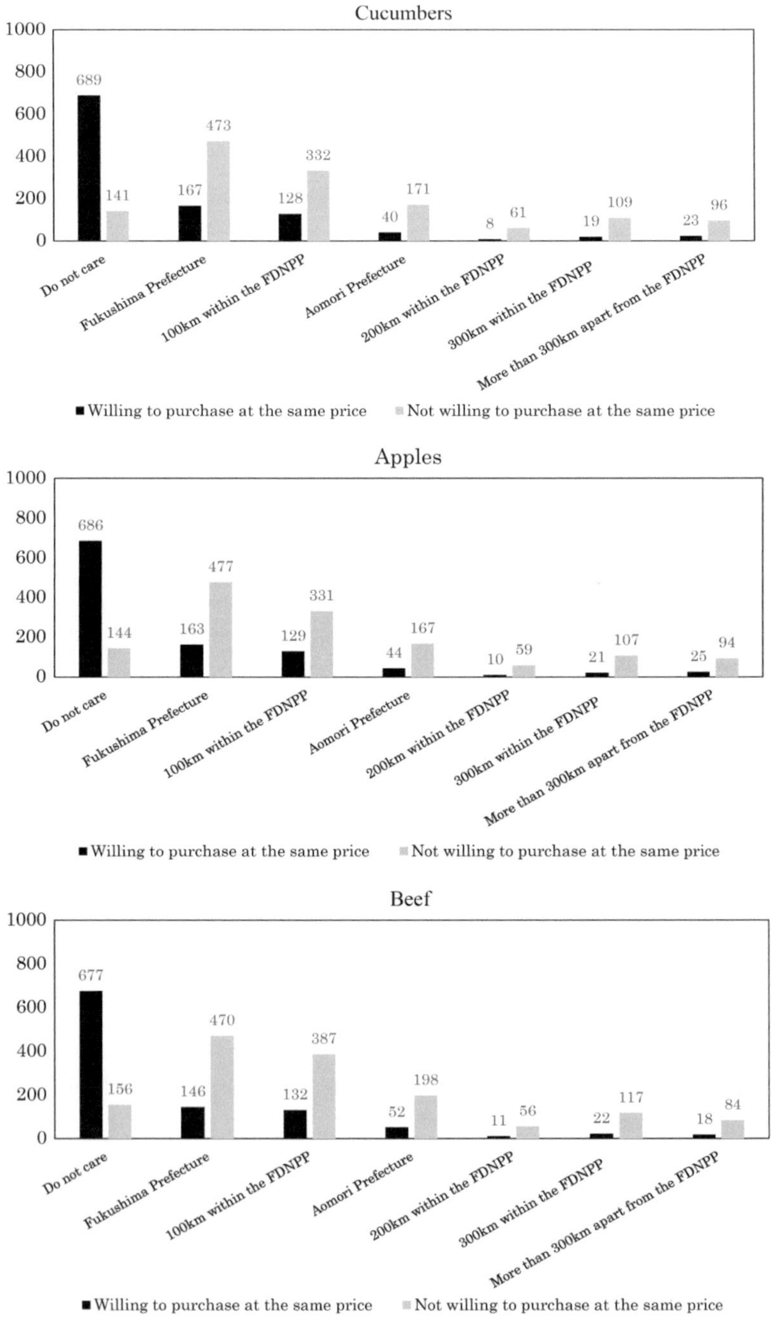

Fig. 5.14 Perceptions toward product origin and the willingness to buy I

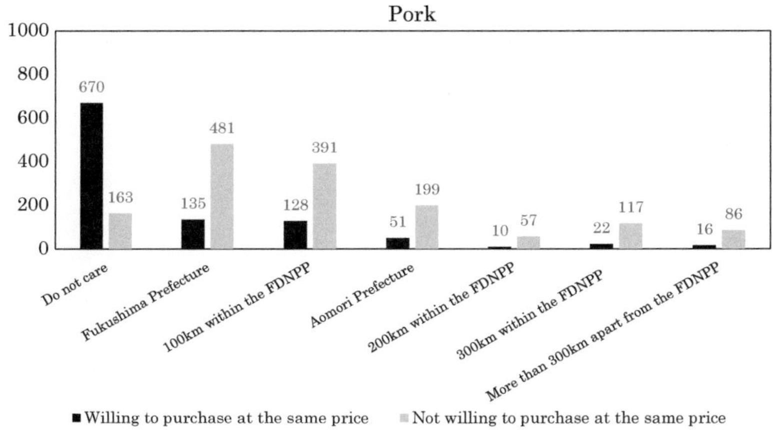

Fig. 5.14 (continued)

Respondents' trust toward food safety standards for radiation and the willingness to buy

In the last part of the subsection, I would like to explain how the difference in the levels of respondents' trust on food safety standards affect their willingness to buy food from regions near the FDNPP.

The "Trust safety standards" in Fig. 5.15 denotes the number of respondents who answered 7 through 10 (very much trust the standard) to the question "Do you trust the current safety standards for radioactive cesium in food?," which is Q15 of Table 5.2. The "Don't' trust safety standards" category belong to respondents who picked 1 (do not trust at all) through 4 to this question.

From Fig. 5.15, it is apparent that people who trust the current Japanese food safety standards for radioactive cesium have positive reaction about buying food from regions near the FDNPP while people who do not trust the safety standards have an adverse reaction. This suggests that improving the reliability of food safety standards for radioactive material will help enhance the sales of food produced in regions near the FDNPP.

5.2.5 Respondents' Perceptions of Willingness to Accept Food from Regions Near the FDNPP and Their Willingness to Buy

Fifth, this subsection will seek how respondents' willingness to accept perceptions for cucumbers, apples, beef, and pork are related to willingness to buy these products from regions near the FDNPP. We will use Table 5.8 and Fig. 5.16 for this purpose.

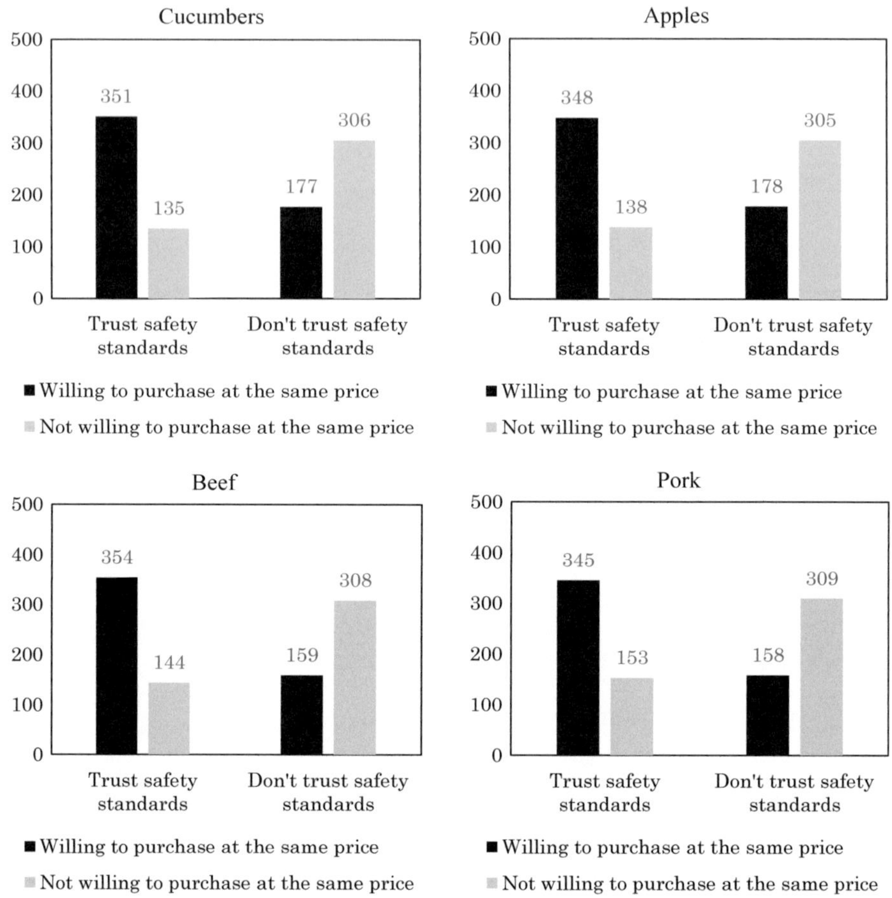

Fig. 5.15 Trust toward food safety standards for radiation and the willingness to buy I

Respondents' willingness to accept perceptions for cucumbers, apples, beef, and pork

Table 5.8 presents the percentages of respondents who are willing to buy cucumbers, apples, beef, and pork produced in regions near the FDNPP at the same price as those from regions farther from the FDNPP, respondents who are willing to buy these products if there is a 10–60% price discount for products from regions near the FDNPP, and finally, respondents who are not willing to purchase products with a 60% price discount. The table is created based on Q16 and Q17 of Table 5.3, and the percentages in the figure are the ratio of number of respondents who belong to one of the three categories to the total number of respondents for each food product.

You can see from Table 5.8 that a majority of the respondents have a positive reaction about buying food from regions near the FDNPP at the same price as that

Table 5.8 Respondent's willingness to accept perceptions for cucumbers, apples, beef, and pork from regions near the FDNPP

	Number of samples	Percentages %
Cucumbers		
Willing to buy at the same price	966	55.0
Willing to buy when the discount rate is between 10 and 60%	318	18.1
Not willing to buy even with a 60% discount rate	473	26.9
Apples		
Willing to buy at the same price	965	54.9
Willing to buy when the discount rate is between 10 and 60%	328	18.7
Not willing to buy even with a 60% discount rate	464	26.4
Beef		
Willing to buy at the same price	940	53.2
Willing to buy when the discount rate is between 10 and 60%	346	19.6
Not willing to buy even with a 60% discount rate	482	27.3
Pork		
Willing to buy at the same price	925	52.3
Willing to buy when the discount rate is between 10 and 60%	351	19.9
Not willing to buy even with a 60% discount rate	492	27.8

from regions apart from the FDNPP in all products. This result is evident because cucumbers, apples, beef, and pork are the products that have higher willingness to buy among the ten food products investigated in this book. However, it is noticeable from the table that more than 25% of the respondents have negative response about buying food from regions near the FDNPP even when there was a 60% price discount for the product.

The effects of having safety labels on food products

Figure 5.16 illustrates the ratio between respondents who answered yes and no to the question asking whether or not the respondents are willing to buy food from regions near the FDNPP if there is a label on the product showing that it has no risk of radioactive contamination. The figure used the answers for Q18 and Q21 of Table 5.3 and the question was only given to the respondents who were not willing to buy food from regions near the FDNPP even when there was a 60% price discount. Thus, the percentages in Fig. 5.16 are among those who responded negatively about buying food from regions near the FDNPP even with a substantial price discount.

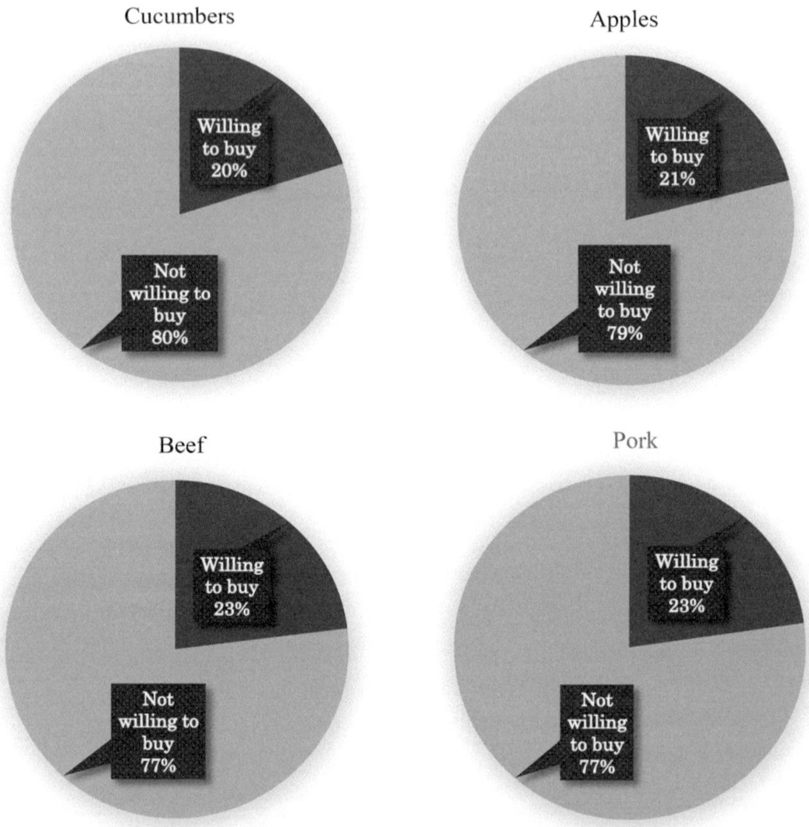

Fig. 5.16 Food labels that shows the product is safe from radiation and the willingness to buy I

According to Fig. 5.16, almost 80% of the respondents within those who refused to buy food from regions near the FDNPP with a 60% price discount continued to show a negative reaction about buying the food even when a label that indicates the product is free from the risk of radioactive contamination is placed on the product. However, this result also means that at least 20% of the respondents do change their response about buying food from regions near the FDNPP if a label certifying the safety of the food from the risk of radioactive contamination is put on the product.

This result implies that having safety labels on food produced in regions near the FDNPP do have positive effects, though small it may be, even on the mind of respondents who consider highly of the risk of radioactive contamination when buying food from these regions.

5.2.6 *Respondents' Social Attributes and Their Willingness to Buy*

Finally, as the sixth part of this section, we will look into how respondents' social attributes have an influence on their willingness to buy cucumbers, apples, beef, and pork from regions near the FDNPP using Fig. 5.17 through Fig. 5.23.

Respondents' residential distance from the FDNPP and the willingness to buy

In Fig. 5.17, we will examine the relationship between the respondents' resident location and the respondents' willingness to buy cucumbers, apples, beef, and pork from regions near the FDNPP. The reason that more respondents who live between 200 and 300 km apart and more than 400 km apart from the FDNPP is because the survey is based on the 2014 population census of Japan and the actual population of Japan is highly concentrated in the Tokyo, Nagoya, and Osaka metropolitan areas.[2] The distance in the figure is estimated from the respondents' zip code of their residence obtained in Q22 of Table 5.4.

From the figures for cucumbers, apples of Fig. 5.17, it is conceivable that in every resident locations more people are willing to buy food from regions near the FDNPP. It is likely that for cucumbers and apples the resident location does not affect the respondents' willingness to buy decisions. Beef too seems to have small difference in the willingness to buy among the respondents' resident locations but the ratio of respondents reacting positively to those reacting negatively about buying beef is smaller compared with cucumbers and apples. This indicates that more people are reacting negatively for beef than cucumbers and apples. For pork, the majority of respondents who live more than 300 km apart from the FDNPP answered not willing to buy pork produced in regions near the FDNPP.

Hence, it is likely that consumers who live apart from the FDNPP seemed to respond negatively about buying food from regions near the FDNPP for beef and pork compared with cucumbers and apples. The reason for beef and pork to be avoided from consumers who live apart from the FDNPP is perhaps because consumers are not used to buying beef and pork from regions near the FDNPP like Fukushima Prefecture because it is uncommon for beef and pork to have the product origin at a prefecture-level. The regulation regarding the labeling of product origins for beef and pork products is not stipulated at a prefecture-level because beef cattle and pigs are often propagated, bred, and slaughtered in different areas and identifying their product origin at a prefecture-level is difficult. This is why consumers rarely see beef and pork products with labels showing that they are from Fukushima Prefecture.

---

[2]Cities with a large population like Tokyo, Yokohama are located 200–300 km apart from the FDNPP. Nagoya and Osaka are located more than 400 km apart from the FDNPP. None of the Tokyo, Nagoya, and Osaka metropolitan areas are located within the 100 km, between 100 and 200 km, and between 300 and 400 km zones from the FDNPP.

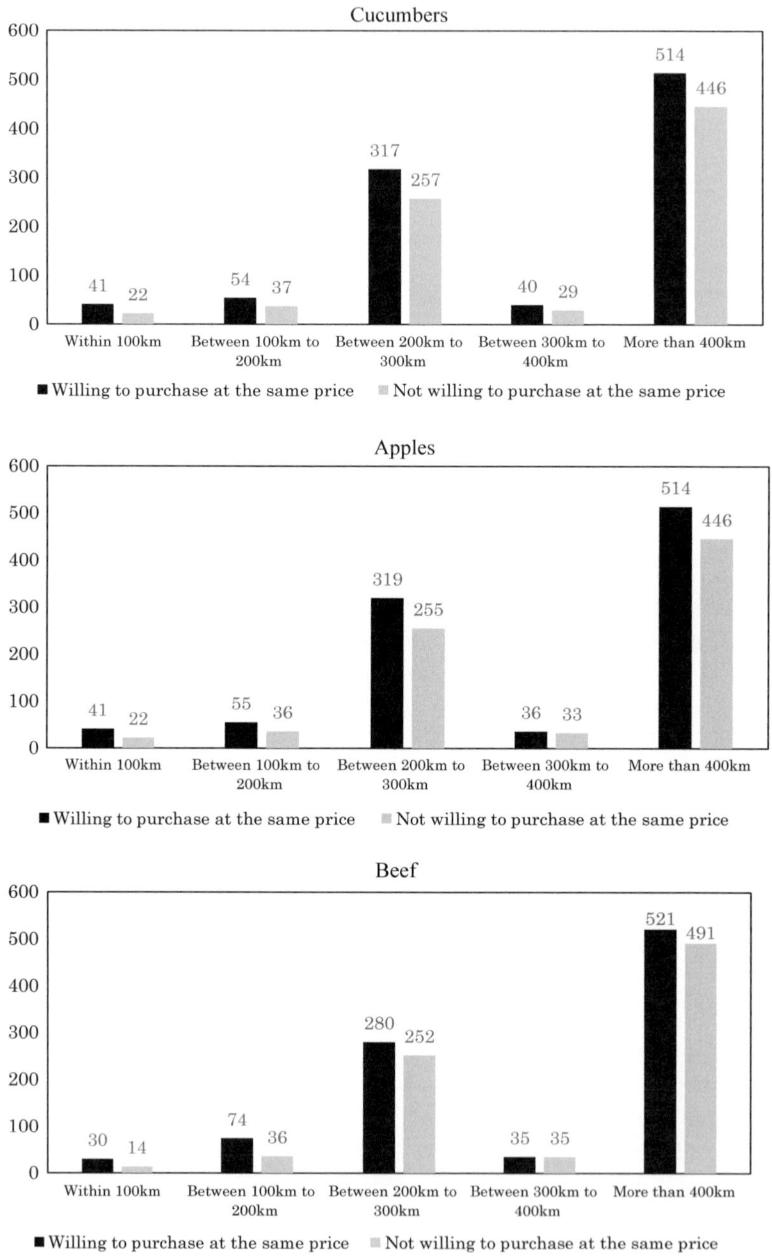

Fig. 5.17 Location of residence and the willingness to buy I

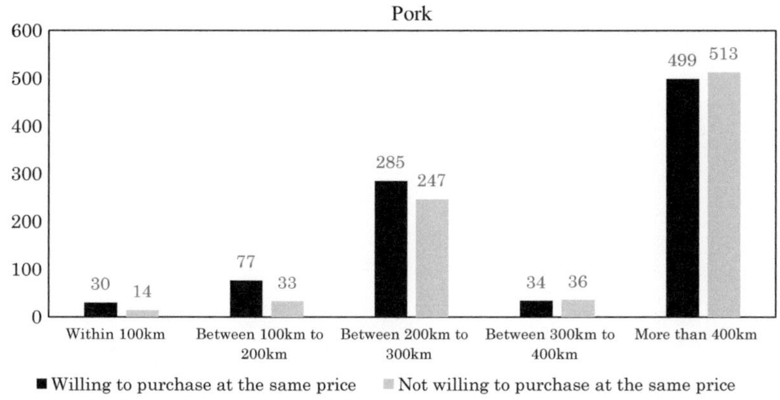

Fig. 5.17 (continued)

On the other hand, the product origin for fruit and vegetables are normally labeled at a prefecture-level. Furthermore, as the share of the Fukushima cucumbers and apples in Japan is relatively large compared with that of beef and pork, they are more accessible in all parts of Japan. Therefore, it is believable that the consumers in Japan are more accustomed to seeing cucumbers and apples from regions near the FDNPP in grocery stores compared with beef and pork. This difference in the familiarity of the product origin might have influenced the willingness to buy of the consumers who live farther from the FDNPP.

Respondents' gender and their willingness to buy

Next, I would like to explain the difference in the willingness to buy cucumbers, apples, beef, and pork from regions near the FDNPP between male and female through Fig. 5.18.

You can see from Fig. 5.18 that male respondents are more likely to accept buying food produced in regions near the FDNPP than female respondents for all food products. The reason for this difference in the willingness to buy between the genders is perhaps because the length of time wives spend with their children is longer than husbands in Japan. This is because the female labor force participation rate of Japan is one of the lowest among the OECD countries and it is still common for women to stay home and take care of her children. Hence, it is likely that female consumers are more worried about the risk of their children getting exposed to radiation after eating food contaminated with radioactive material, and for this reason, they want to avoid buying food from regions near the FDNPP.

Respondents' age and their willingness to buy

Figure 5.19 exhibits the relationship between the respondents' age and their willingness to buy cucumbers, apples, beef, and pork from regions near the FDNPP.

You can see from the figure that respondents who are above 50 years of age are more likely to accept buying food produced in regions near the FDNPP while

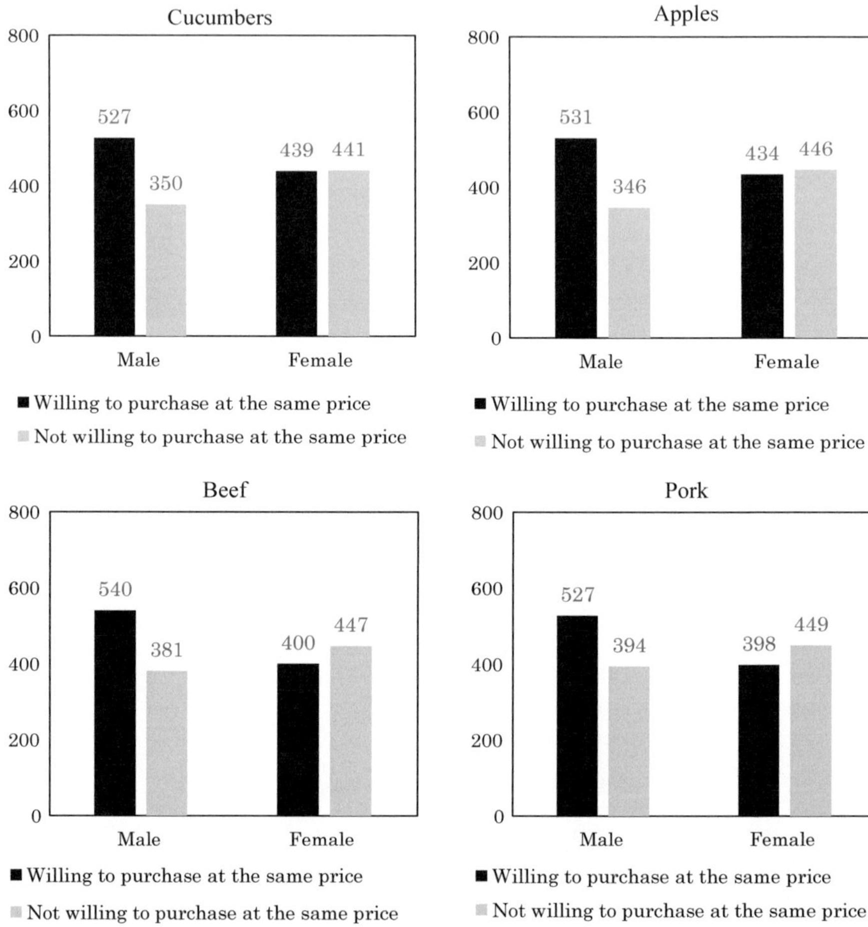

Fig. 5.18 Gender and the willingness to buy I

respondents whose age are below 50 have a large number of respondents who react negatively about buying food from such regions. Thus there seems to be a tendency that younger people are more likely to avoid buying food from regions near the FDNPP.

In particular, respondents in the age 30s have the highest number of people who are not willing to buy food from regions near the FDNPP among all the age groups in the figure. The reason for this avoiding behavior among the respondents of age 30s is probably because the rate of having children is high among these respondents and these respondents are concerned about the risk of their children getting exposed to radiation.

Cucinotta and Durante (2009) suggests that the risk of becoming cancer when exposed to radiation is higher for younger children and it is likely that more people

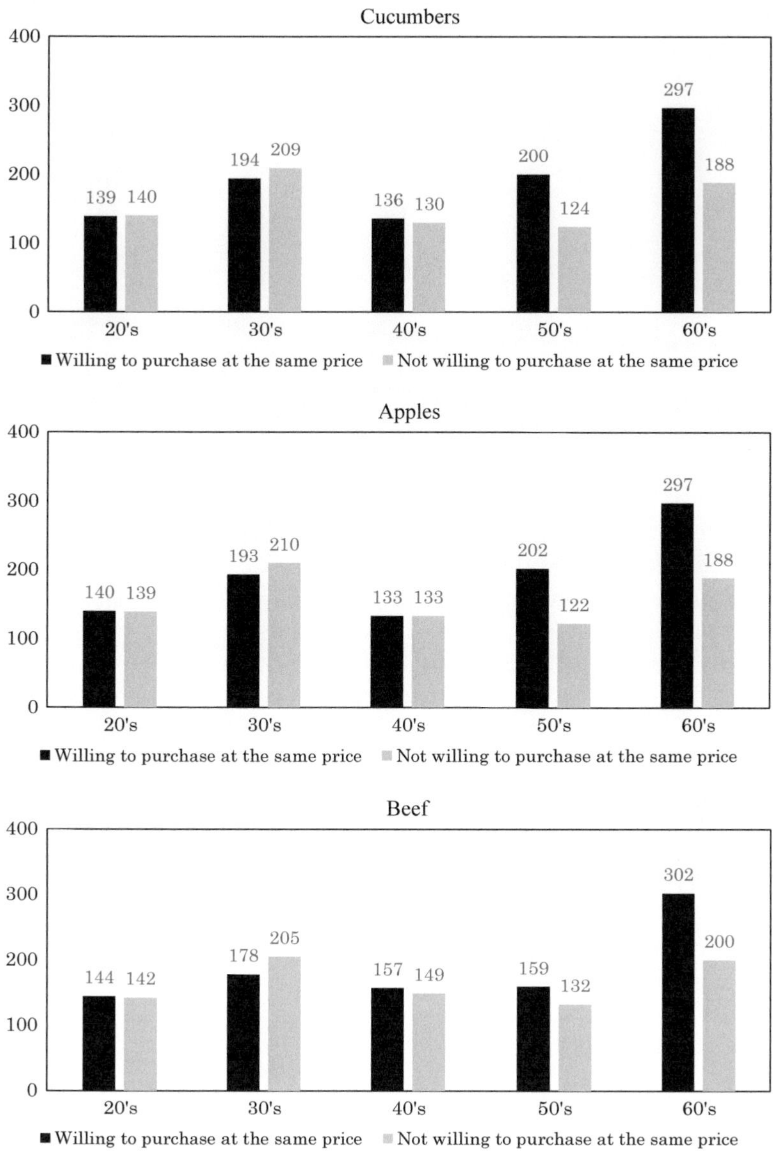

Fig. 5.19 Age and the willingness to buy I

became aware of this information after the Fukushima disaster because such risk has often been broadcasted and spread by media after 2011. Hence, it could be that this spread of information related to children and radiation might have influenced the results in Fig. 5.19.

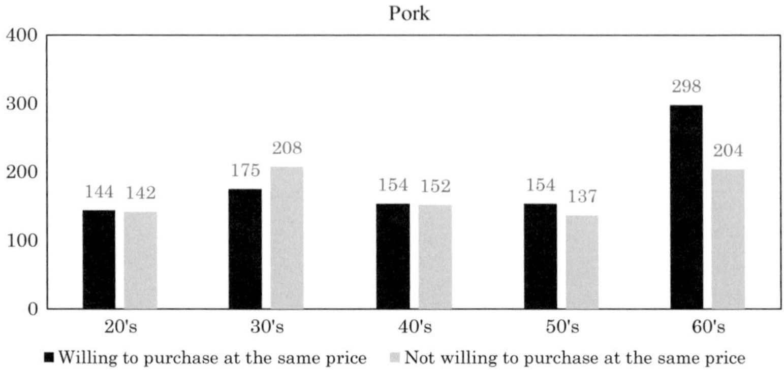

Fig. 5.19 (continued)

Relationships between the presence or absence of children and the willingness to buy

Now I would like to explain how the presence or absence of children in the household of respondents have effects on the respondents' willingness to buy cucumbers, apples, beef, and pork produced in regions near the FNDPP.

Figure 5.20 shows the relationship between the respondents' family member structure and the respondents' willingness to buy food from regions near the FDNPP. The figure is created by using Q25 of Table 5.4. As seen in the figure, respondents who have children under the age of elementary school in their household have a higher number of respondents that have a negative response about buying food from regions near the FDNPP. On the other hand, it is apparent that respondents who have a family member of high school age are more likely to buy food from regions near the FDNPP and those who have a family member that is over 65 of age have an even higher likelihood of buying food from these regions.

Similarly, you can see in Fig. 5.21 that respondents show negative reaction about buying food from regions near the FDNPP when they live with more than two children that are age under 15. Figure 5.21 depicts the relationship between the number of children under age 15 in respondents' household and the respondents' willingness to buy food from regions near the FDNPP. The figure is based on the answers of Q26 of Table 5.4. It is also noticeable from the figure that respondents with only one child in their household have a positive willingness to buy. Many elderly people share their house with their children to help raise their grandchildren when little in Japan. Therefore, we believe the reason why a relatively large number of respondents with one child answered positively about buying food near the FDNPP is because there are respondents who have only one child in their household, but do not have meals together.

All in all, it is conceivable that presence of children in the household has a negative impact on the willingness to buy cucumbers, apples, beef, and pork produced in regions near the FDNPP.

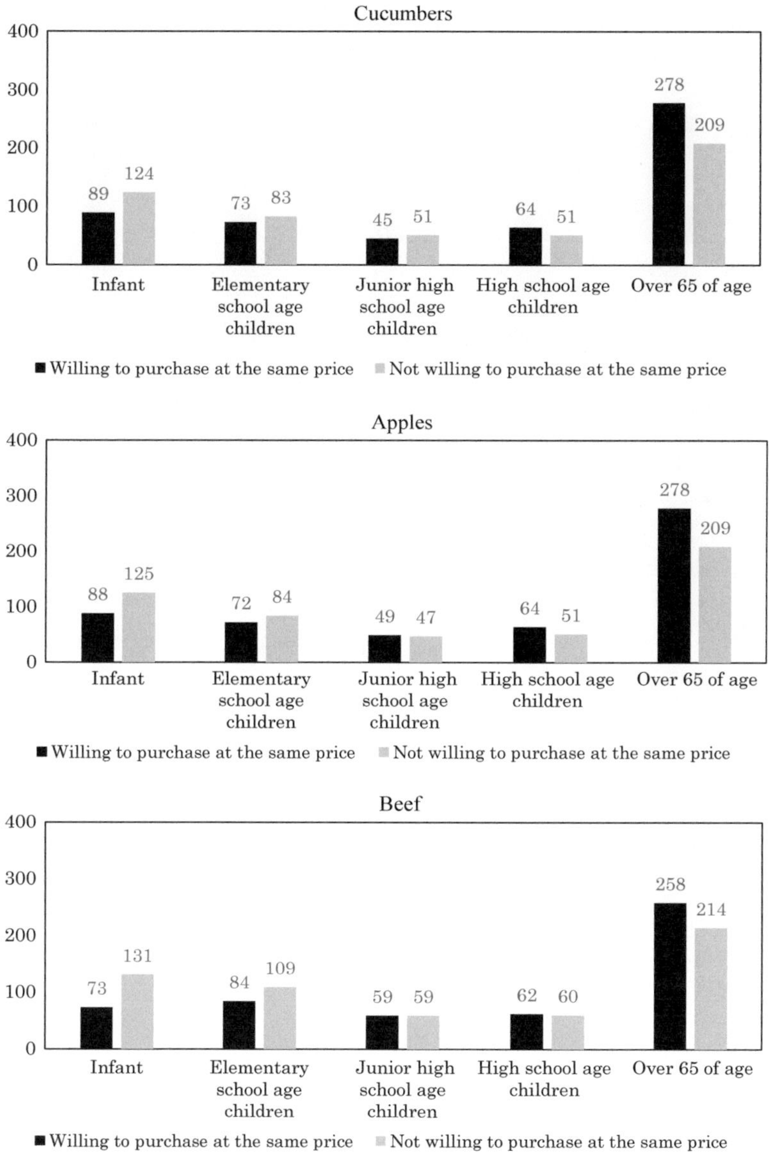

Fig. 5.20 Family member structure and the willingness to buy I

Educational level and the willingness to buy

Figure 5.22 illustrates the relationship between the respondents' highest level of education attained and the willingness to buy cucumbers, apples, beef, and pork from regions near the FDNPP. The category "High school" consists of respondents

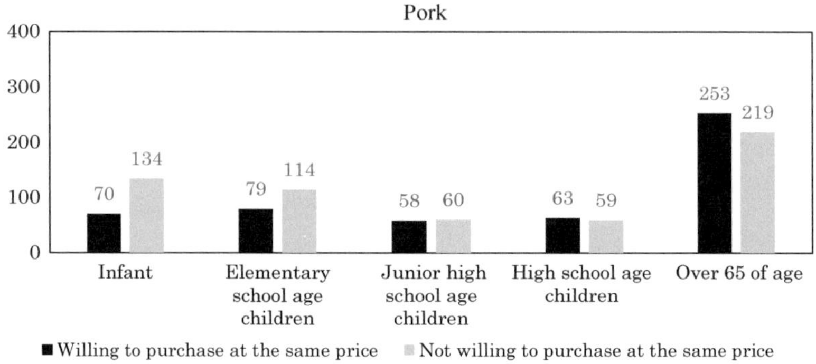

Fig. 5.20 (continued)

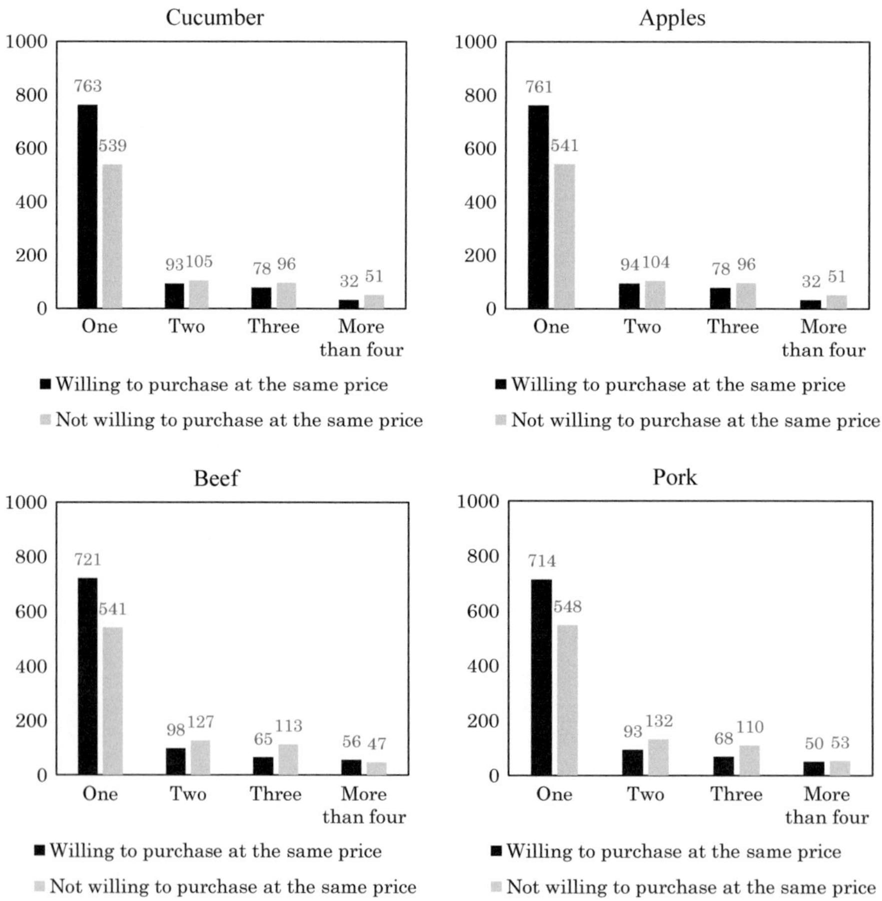

Fig. 5.21 Number of children under age 15 and the willingness I

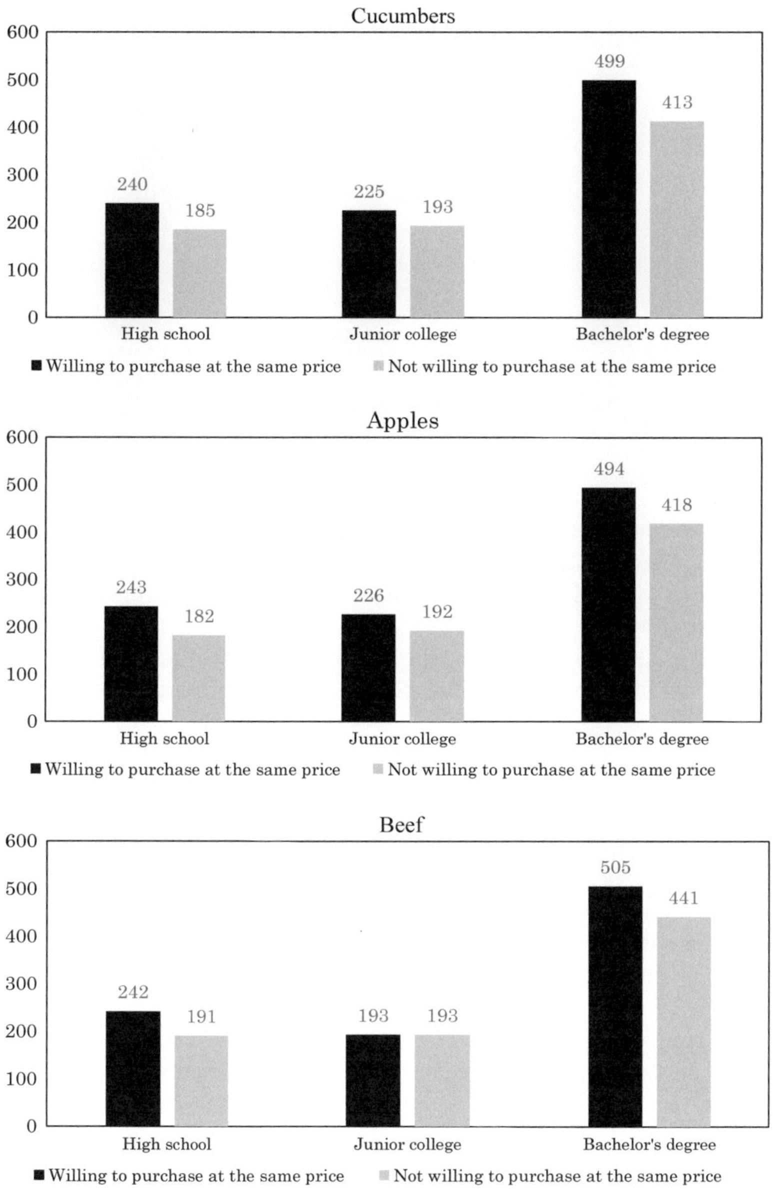

Fig. 5.22 Highest educational level and the willingness to buy I

selecting 1 and 2 in Q27 of Table 5.4, "Junior college" consists of those choosing 3 and 4 in Q27, and "Bachelor's degree" denotes those who picked 5, 6, and 7 in Q27 achieving at least a bachelor's degree.

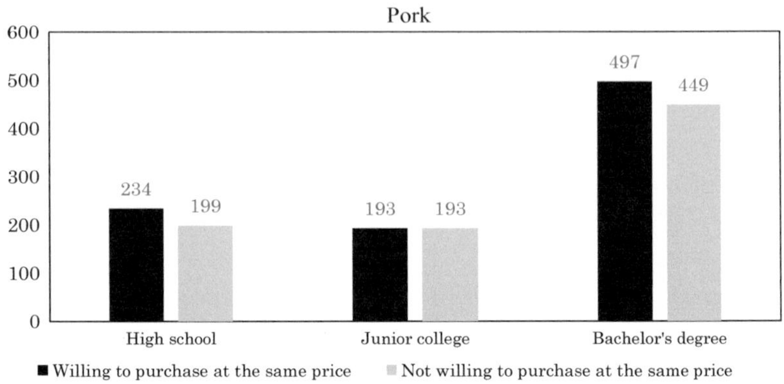

Fig. 5.22 (continued)

From Fig. 5.22 it is conceivable to assume that in all food products more respondents accept buying food from regions near the FDNPP at the same price as those living farther from the FDNPP. The ratio of respondents "willing to purchase" to "not willing to purchase" seems to decline among respondents with a junior college degree, but it increases again among those with a bachelor's degree. Therefore, it is likely that the level of education does not have an effect on the respondents' willingness to buy cucumbers, apples, beef, and pork from regions near the FDNPP.

Respondents' income level and the willingness to buy

Finally, as the last component of this section, I would like to introduce the relationship between the respondents' annual income and the willingness to buy cucumbers, apples, beef, and pork from regions near the FDNPP using Fig. 5.23. The figure is formed with the answers obtained in Q27 of Table 5.4. The samples for respondents with an annual income of 10 million yen was very slight in the survey so they are included in the respondents with annual income above 8 million yen.

It can be seen from Fig. 5.23 that in most of the income categories, a majority of the respondents have a positive reaction about buying cucumbers, apples, and beef from regions near the FDNPP at the same price as regions apart from the FDNPP. By contrast, for pork, there are nearly the same number of respondents who reacted negatively about buying pork from regions near the FDNPP among those who have annual income of fewer than 2 million yen and more than or equal to 8 million yen.

First the reason for having a high number of respondents avoiding pork from regions near the FDNPP among those with annual income of more than 8 million yen is perhaps because paying a higher price to avoid buying pork from regions near the FDNPP is not a burden for people with high income. Hence, it is likely that

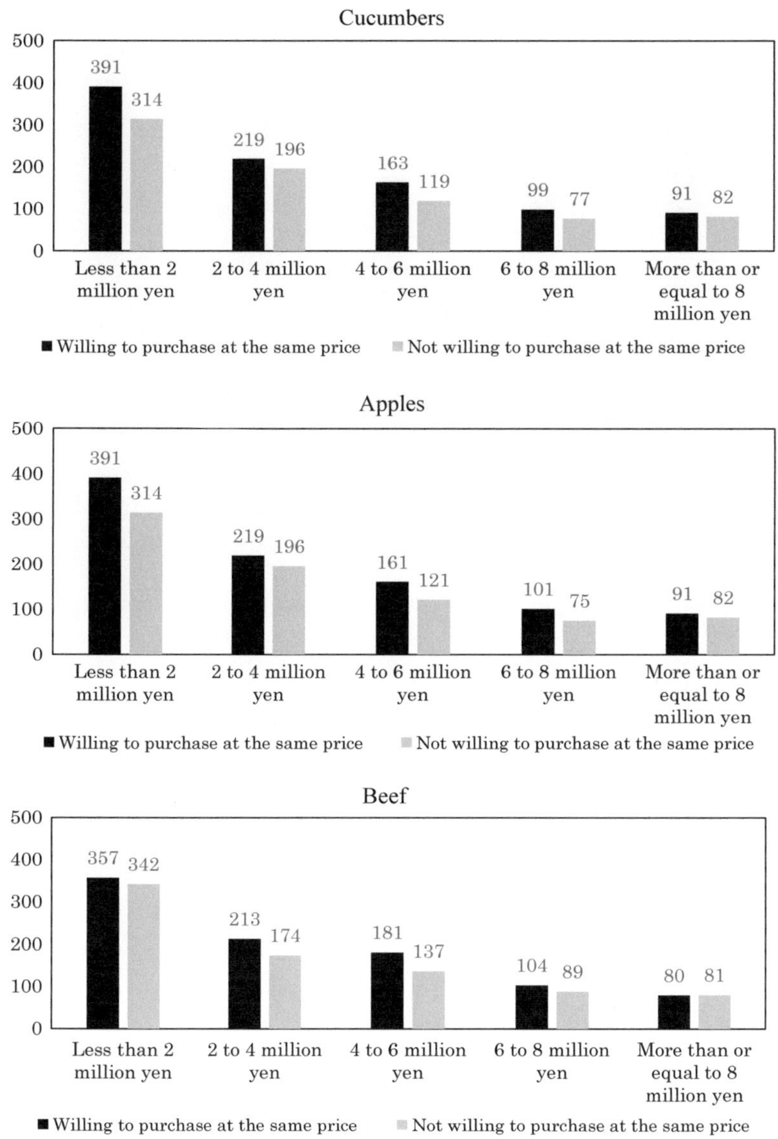

Fig. 5.23 Income level and the willingness to buy I

even having a discount in the price of pork from regions near the FDNPP might be meaningless for such people because they are generally willing to pay a higher price to avoid any risk involved in pork produced in regions near the FDNPP.

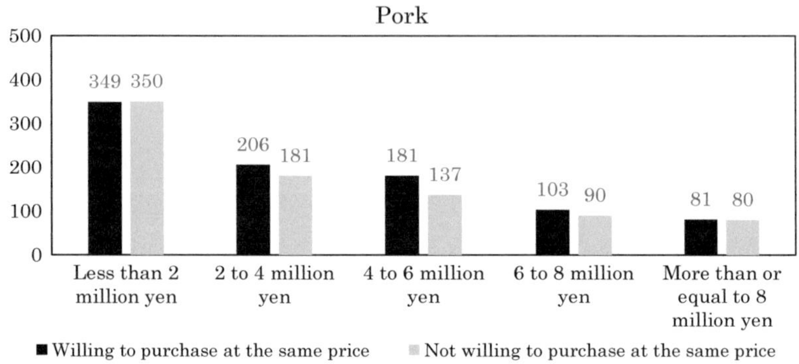

Fig. 5.23 (continued)

Second, the reason for the respondents who have an annual income of less than 2 million yen to avoid pork from regions near the FDNPP is related to the percentages of female respondents that belong to this income category. We find that 73% of the respondents who answered the survey for pork and have annual income below 2 million yen are women. There are quite a few women who are not fully employed and work as housewives to take care of their children in Japan. These women are those who prepares the meal in their household so that they are more aware of the risk of their children getting exposed to radiation from having food contaminated with radioactive material. It could be that such housewives who are in the group of an annual income of below 2 million yen affected the results of the response for willingness to buy pork from regions near the FDNPP.

5.3 Food Produced in Regions Near the FDNPP that About a Half of the Respondents Are Willing to Buy: Shiitake Mushrooms, Chicken Eggs, and Tuna Fish

In this section, we investigate what factors might affect the consumers' willingness to buy toward shiitake mushrooms, chicken eggs, and tuna fish produced in regions near the FDNPP, which are foods that about a half of the respondents show a positive willingness to buy. Here too, we will ascertain how the respondents' eating habits, perceptions of food safety, interests in social problems, attitudes toward nuclear contamination, consumers' willingness to accept food produced near the FDNPP, and social attributes are related to the consumers' willingness to buy shiitake mushrooms, chicken eggs, and tuna fish from these nearby regions.

5.3.1 *Respondents' Eating Habits and Their Willingness to Buy*

Similarly, as in the previous section, first we will examine how respondents' eating habits have an influence on the respondents' willingness to buy shiitake mushrooms, chicken eggs, and tuna fish produced in regions near the FDNPP. For this purpose, we will go over the results in Fig. 5.24 through Fig. 5.28 in this subsection.

Shopping frequency and willingness to buy

Figure 5.24 depicts how often the respondents' go shopping to buy grocery and how such shopping frequency is related to the respondents' willingness to buy shiitake mushrooms, chicken eggs, and tuna fish from regions near the FDNPP.

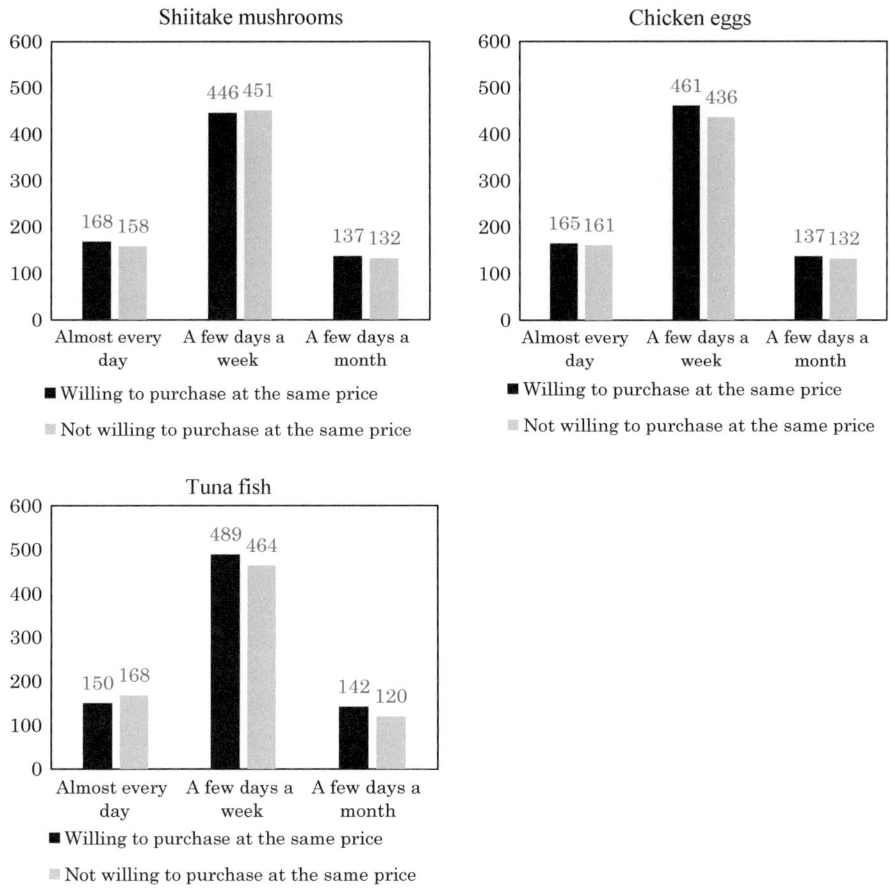

Fig. 5.24 Shopping frequency and the willingness to buy II

From the figure, you can see that there is not much difference in the respondents' willingness to buy among the respondents' shopping frequency for shiitake mushrooms and chicken eggs. In both foods, whether the respondents go shopping almost every day or only a few days a month, there are not much difference between the number of respondents reacting positively and negatively about buying food from regions near the FDNPP.

On the other hand, for tuna fish, there is a difference in the respondents' willingness to buy between the respondents that go shopping almost every day and only a few days a month. It is apparent from the figure that respondents of high shopping frequency are more likely to avoid buying tuna fish from regions near the FDNPP. This might be suggesting that consumers who often go shopping are more concerned about the risk of radioactive contamination for tuna fish produced in regions near the FNDPP.

Purchase experience and the willingness to buy

Next in Fig. 5.25 we will seek the relationship between the consumers' purchase experience of shiitake mushrooms, chicken eggs, and tuna fish and consumer's willingness to buy these foods from regions near the FDNPP.

Within the respondents who have purchased shiitake mushrooms, chicken eggs, and tuna fish for themselves, more individuals are willing to buy these products from regions near the FDNPP. Compared with this chart, fewer people are willing to buy these products if they are from regions near the FDNPP if the respondents never purchased such foods.

This result is different from the case of cucumbers, apples, beef, and pork shown in Fig. 5.4 where purchase experience did not have effects on the respondents' willingness to buy these products from regions near the FDNPP. The reason of the difference is perhaps because individuals with no experience in buying shiitake mushrooms, chicken eggs, and tuna fish reacted negatively about buying these products from regions near the FDNPP. It is presumable that such consumers are more likely to perceive highly of the risk of radioactive contamination in foods.

Cooking frequency and the willingness to buy

Figure 5.26 elucidates how the difference in the respondents' cooking frequency is related to the respondents' willingness to buy shiitake mushrooms, chicken eggs, and tuna fish from regions near the FDNPP.

You can see from the figure that among the respondents who prepare meals almost every day for themselves or their family, more respondents want to avoid buying the three foods in the figure if they are from regions near the FDNPP. For tuna fish, respondents have a negative response even among those who cook meals a few days a week. In contrast, for shiitake mushrooms and chicken eggs, more respondents answered that they are willing to buy food from regions near the FDNPP among those who cook meals a few days a week or a month. Hence, it is likely that consumers who frequently cook meals have a negative feeling about buying tuna fish from regions near the FDNPP.

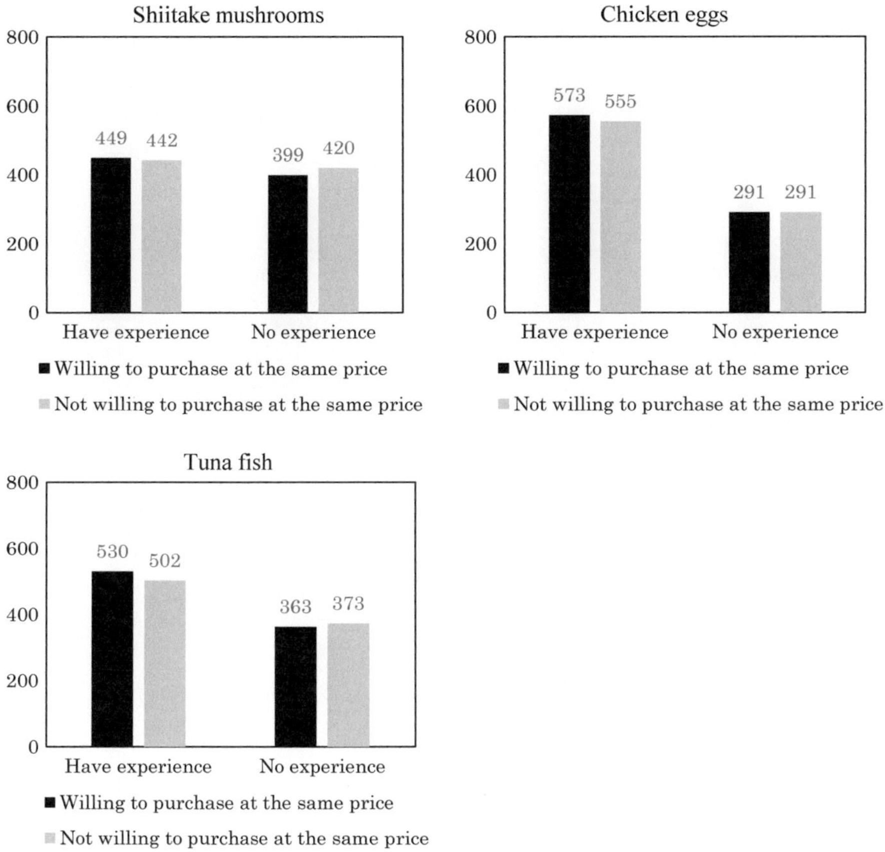

Fig. 5.25 Purchase experience and the willingness to buy II

Among the respondents who do not cook meals so often, it is apparent from the figure that more people are willing to buy shiitake mushrooms, chicken eggs, and tuna fish from regions near the FDNPP compared with those who never cook meals for themselves or their family. The reason for having a higher avoiding behavior among the consumers who never cook meals is probably because these consumers do not buy the foods in the figure for themselves. As I previously explained in Fig. 5.25, individuals who do not buy food for themselves are more likely to belong to a group of individuals who never cook meals for themselves or for their family. In fact, among the respondents who answered that they never cook meals, 85, 78, and 82% of the respondents in the survey did not have experience in purchasing shiitake mushrooms, chicken eggs, and tuna fish for themselves, respectively. Thus, people who never cook meals hardly buy shiitake mushrooms, chicken eggs, and tuna fish for themselves and it is likely that this inexperience in both cooking and

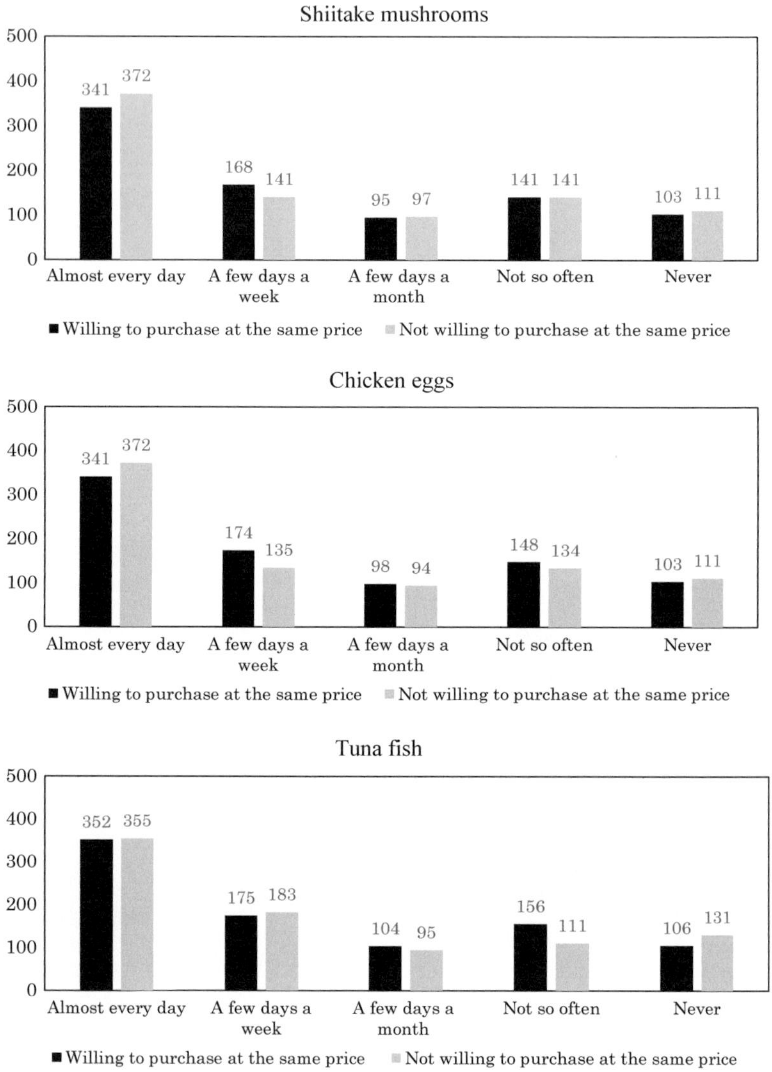

Fig. 5.26 Cooking frequency and the willingness to buy II

purchasing food is affecting the respondents to react negatively about buying these foods from regions near the FDNPP.

Eating style and the willingness to buy

Figure 5.27 describes how differences in the way respondents take meals are related to the consumers' willingness to buy shiitake mushrooms, chicken eggs, and tuna fish from regions near the FDNPP.

From the figure, more than half of the respondents are not willing to buy shiitake mushrooms and chicken eggs from regions near the FDNPP suggesting that consumers who mostly eat their meals at home are more likely to respond negatively about buying shiitake mushrooms and chicken eggs from such regions. On the contrary, within the respondents who mostly go out for their meals, more of them answered positively about buying shiitake mushrooms and chicken eggs from regions near the FDNPP. This difference in the response for shiitake mushrooms and chicken eggs between the respondents' eating style is perhaps because not all restaurants in Japan provide customers with all the product origins for the ingredients each dish served. Therefore, the respondents who often go out to eat are less sensitive about the product origin for shiitake mushrooms and chicken eggs that are included in their meals.

However, the result for the tuna fish in Fig. 5.27 has a different outcome. As seen in the figure for tuna fish, even respondents who eat at home have a positive

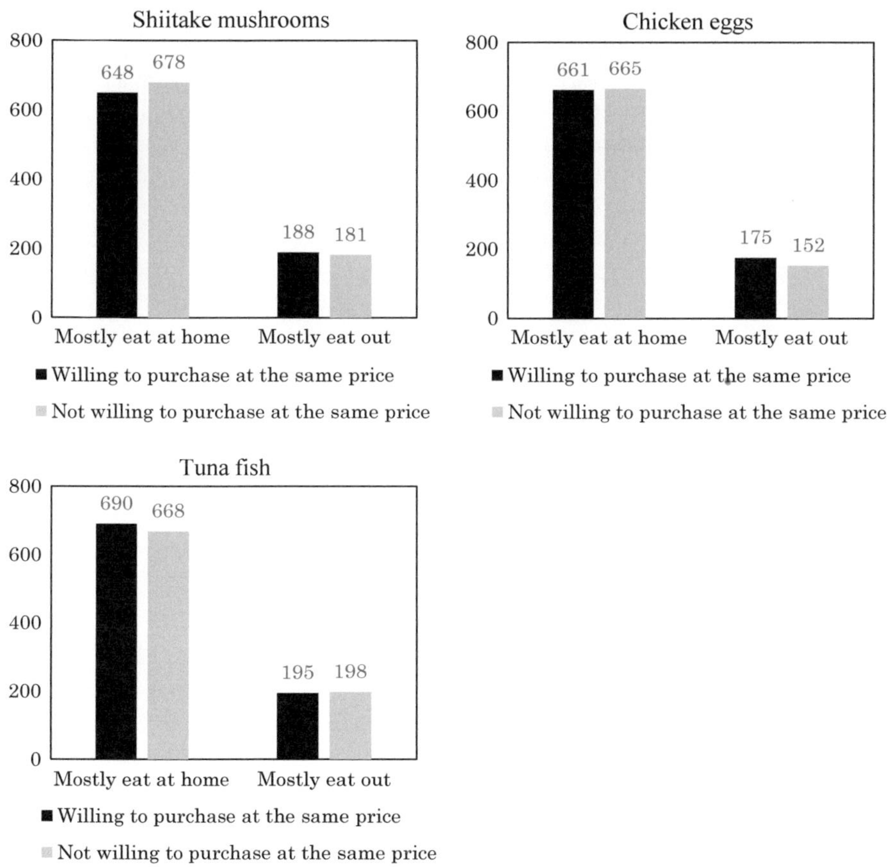

Fig. 5.27 Eating style and the willingness to buy II

reaction about buying tuna fish from regions near the FDNPP. This is likely related to the low frequency of having tuna fish at home compared with shiitake mushrooms and chicken eggs. As people do not cook or buy tuna fish so often, it is likely that the respondents have the tendency to evaluate the risk of purchasing radioactively contaminated tuna fish to be lower compared with shiitake mushrooms and chicken eggs.

Respondents' concerns when buying food and the willingness to buy

As the last part of this subsection, I would like to show the relationship between the factors that consumers worry the most when buying food products and the consumers' response toward willingness to buy shiitake mushrooms, chicken eggs, and tuna fish from regions near the FDNPP.

It is apparent from Fig. 5.28 that people who are concerned about the quality, freshness, expiration and best buy dates, and safety are more likely to respond negatively about buying shiitake mushrooms and chicken eggs from regions near the FDNPP. By contrast, among the respondents that care the price of a product the most when buying shiitake mushrooms and chicken eggs, it is apparent from the figure that more respondents are willing to buy them from regions near the FDNPP.

This trend of respondents who put importance on price when buying a grocery to have a positive willingness to buy also applies to tuna fish harvested in regions near the FDNPP. Hence, if the regions near the FDNPP can appeal the attractiveness of their price for shiitake mushrooms, chicken eggs, and tuna fish, it could help improve the willingness to buy for these foods produced in these regions.

For tuna fish, it is noticeable from the figure that respondents who picked other than safety are all showing a positive willingness to buy which implies that providing good quality and freshness might help increase the sales of tuna fish from regions near the FDNPP. However, consumers who put importance on safety when buying a grocery are more likely to avoid tuna fish from these regions so advertising the safety of their tuna fish is an important task that needs to be addressed in these regions.

5.3.2 Respondents' Perceptions of Food Safety Issue and Their Willingness to Buy

It is conceivable that consumers' perceptions of food safety issue are one of the important factors that affects the consumers' willingness to buy food from regions near the FDNPP. Thus, in this subsection, I would like to examine how consumers' perceptions of food safety issue is related to the consumers' willingness to buy shiitake mushrooms, chicken eggs, and tuna fish. These are foods that about a half of the respondents in the survey agreed to buy even if they are produced in regions near the FDNPP.

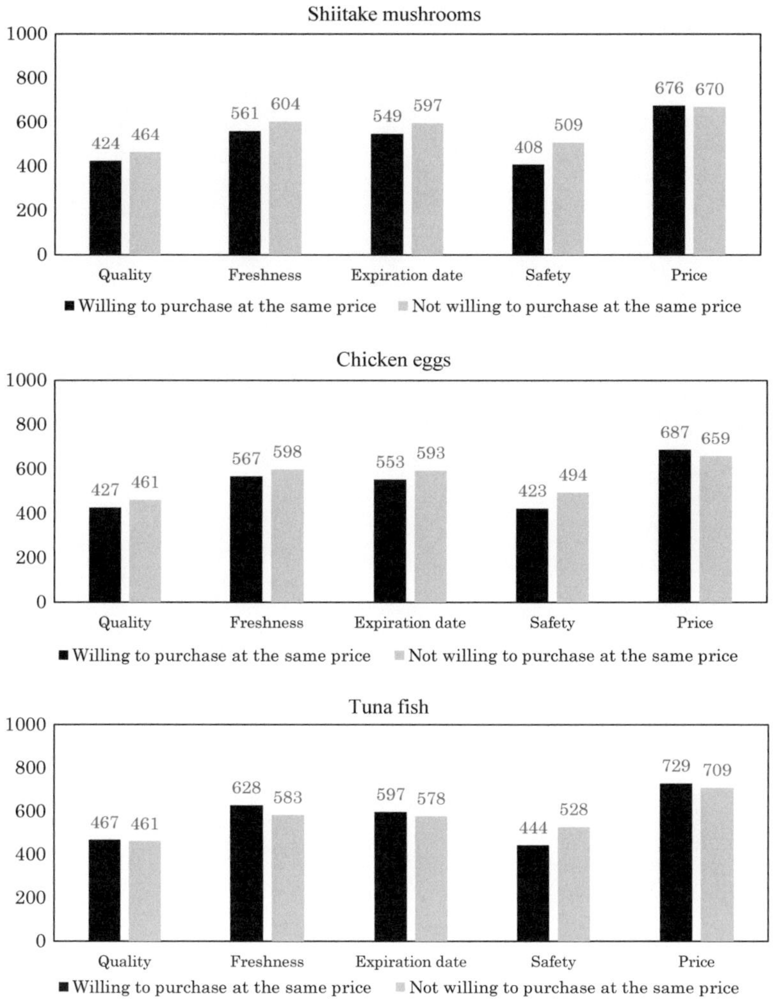

Fig. 5.28 Factors that consumers worry about when buying food and the willingness to buy II

Respondents' perceptions of unsafe food and the willingness to buy

Figure 5.29 presents how consumers' actions to keep themselves away from unsafe food are related to consumers' willingness to buy shiitake mushrooms, chicken eggs, and tuna fish from regions near the FDNPP.

You can see from the figure that among the respondents who selected any of the choices besides "Do not do anything special" are more likely to avoid food from regions near the FDNPP. Consequently, this suggests that respondents who do have safety concerns when buying food products are more worried about the risk of radioactive contamination involved in food from these regions. On the other hand,

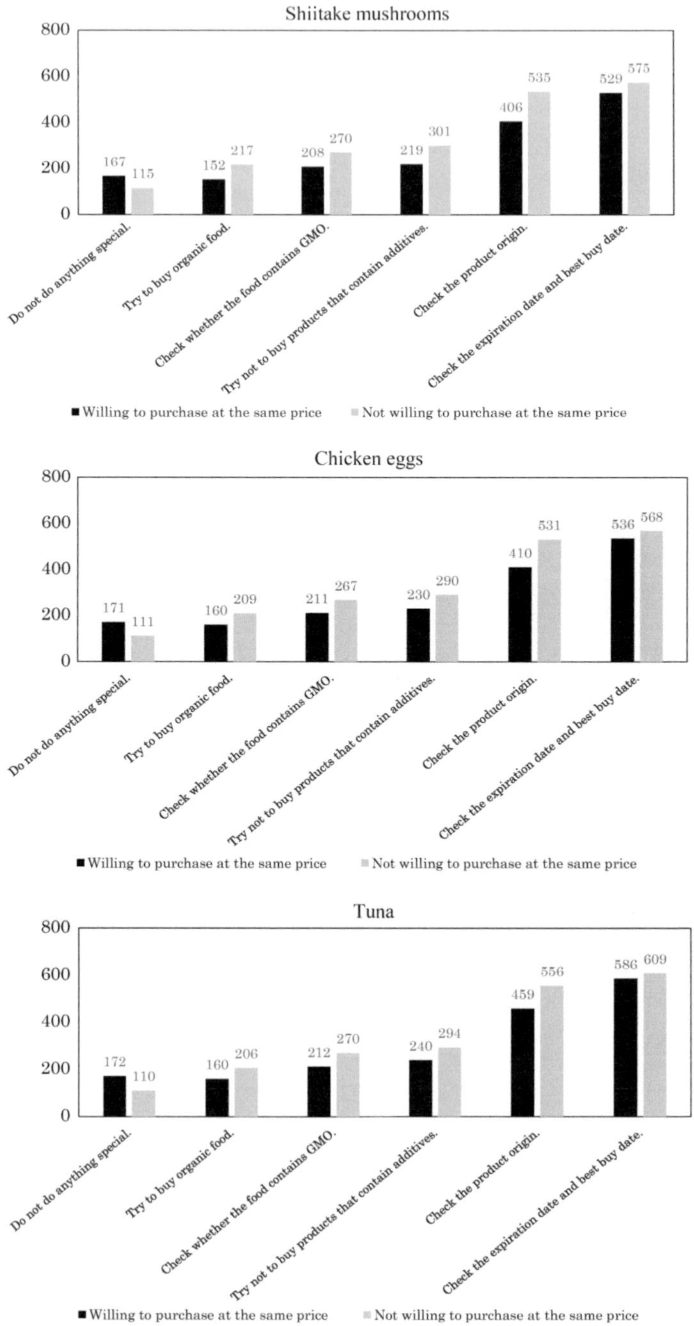

Fig. 5.29 Consumers' actions to keep themselves away from unsafe food and the willingness to buy II

respondents who answered that they "Do not do anything special" to keep themselves away from unsafe food have a positive willingness to buy for food produced in regions near the FDNPP.

These results imply that consumers who perceive the food safety issue very seriously are more reluctant to buy shiitake mushrooms, chicken eggs, and tuna fish from regions near the FDNPP.

Trust in food labels and the willingness to buy

Next, we will seek in Fig. 5.30 how the difference in the levels of trust the respondents put on the information provided in food labels is related to the respondents' willingness to buy shiitake mushrooms, chicken eggs, and tuna fish from regions near the FDNPP.

From Fig. 5.30 it is discernible that there is a gap between respondents who trust the information in food labels and those who do not. Respondents who trust the information in food labels are more likely to react positively about buying shiitake mushrooms, chicken eggs, and tuna fish produced in regions near the FDNPP but

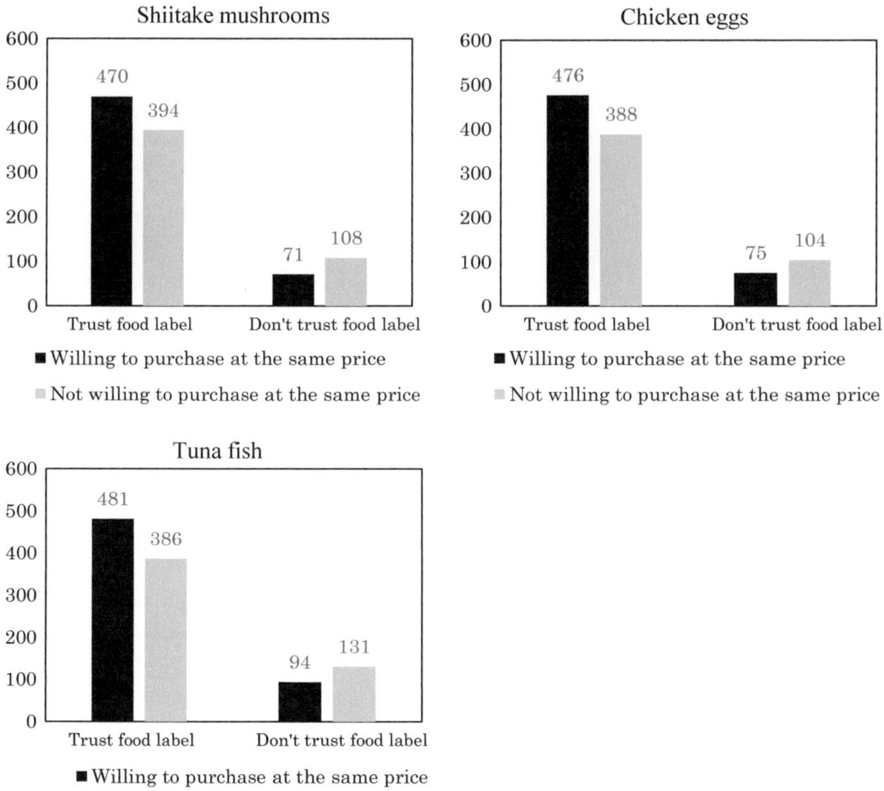

Fig. 5.30 Trust in food labels and the willingness to buy II

those who do not trust the food labels have a tendency to avoid buying food from regions near the FDNPP.

As I mentioned beforehand when I discussed the relationship between confidence in food labels and the willingness to buy for cucumbers, apples, beef, and pork, enhancing the reliability of information the consumers receive from food labels is the key factor in enabling them to purchase shiitake mushrooms, chicken eggs, and tuna fish products from regions near the FDNPP. By doing so, will help gain credibility from the consumers.

5.3.3 Respondents' Interests in Social Problems and the Willingness to Buy

In this subsection, as the third subsection of the section, we will investigate how consumers' interests in social problems have an influence on the consumers' willingness to buy shiitake mushrooms, chicken eggs, and tuna fish from regions near the FDNPP. For this objective, we will look into how the consumers' perceptions of environmental conservation and helping the disaster regions are related to the consumers' willingness to buy shiitake mushrooms, chicken eggs, and tuna fish produced in regions near the FDNPP.

Respondents' perceptions of environmental conservation and their willingness to buy

First, I would like to explain whether consumers who are ardent about environmental conservation are more willing to buy food from regions near the FDNPP. From Fig. 5.31, it is evident that respondents who are more willing to attend activities to conserve the environment are more likely to buy shiitake mushrooms, chicken eggs, and tuna fish from regions near the FDNPP. By contrast, as seen in the figure, respondents who are not so enthusiastic about participating in activities to help conserve the environment are less willing to buy shiitake mushrooms, chicken eggs, and tuna fish produced in regions near the FDNPP.

Respondents' perceptions of helping the disaster regions and their willingness to buy

Figure 5.32 shows how consumers' eagerness to support the disaster regions from the Tohoku-Pacific Ocean Earthquake is related to consumers' willing to buy shiitake mushrooms, chicken eggs, and tuna fish from regions near the FDNPP. Here too, as we saw in the previous subsection, respondents who are more willing to support the disaster regions to recover from their damages are more likely to accept buying food from regions near the FDNPP. On the other hand, respondents that are reluctant about helping the disaster regions have a negative reaction about buying food from such regions.

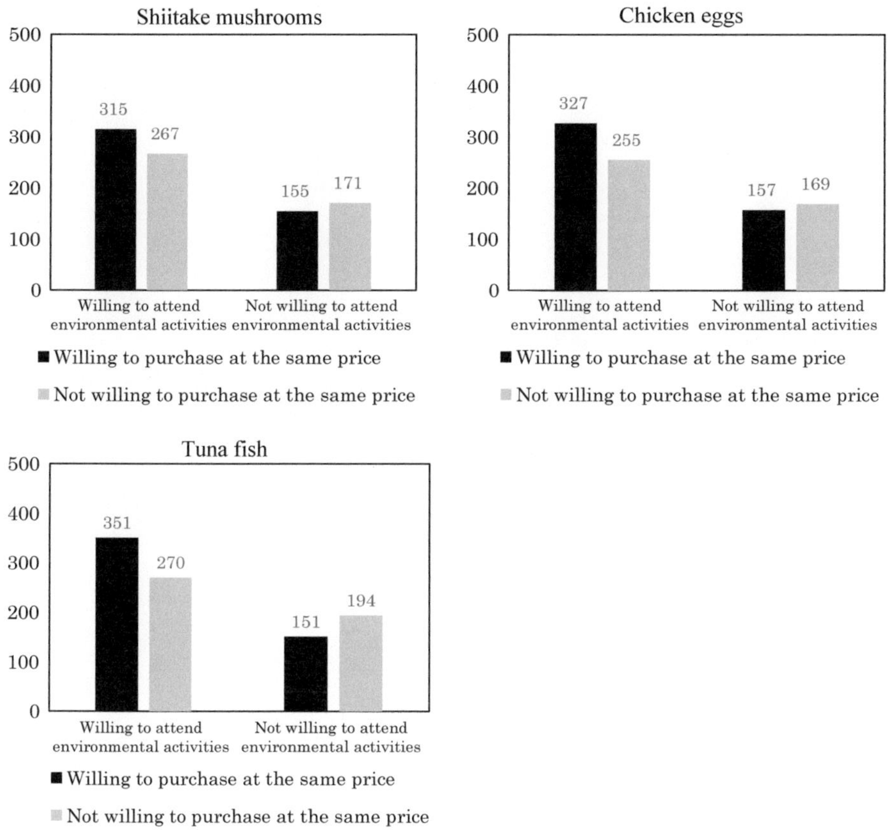

Fig. 5.31 Aggressiveness in preserving the environment and the willingness to buy II

From the above analyses of Figs. 5.31 and 5.32, consumers who are more interested in help improving the environment and conditions of the disaster areas of the Tohoku-Pacific Ocean Earthquake are more positive about buying shiitake mushrooms, chicken eggs, and tuna fish from regions near the FDNPP. This result is similar to our findings in the foods we examined in the previous subsection and it is likely implying that people who are more interested in social problems are more willing to buy food from regions near the FDNPP. It is presumable that this is because such consumers are altruistic consumers thinking that their consumption behaviors will help reinvigorate the economies of the regions near the FDNPP.

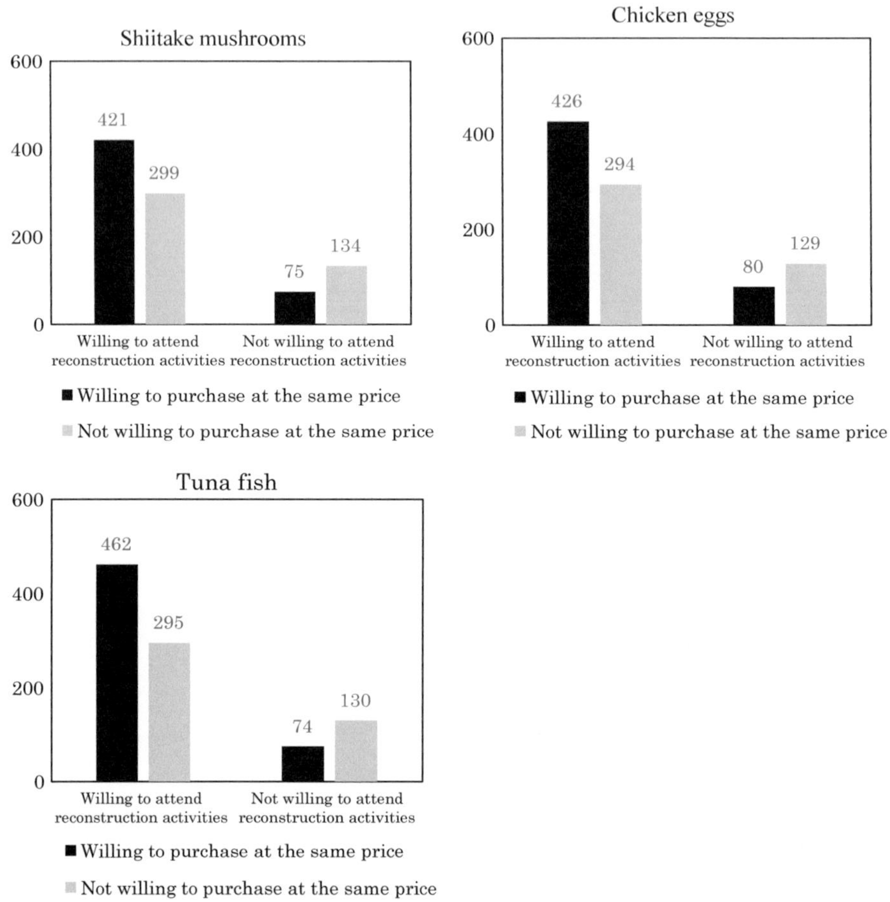

Fig. 5.32 Aggressiveness in reconstructing the regions suffered from the disaster and the willingness to buy II

5.3.4 Respondents' Perceptions of Radioactive Contamination and Their Willingness to Buy

Respondents' risk perceptions toward radioactively contaminated food and their willingness to buy

As the fourth subsection of this section, we will examine the relationship between consumers' perceptions of radioactive contamination and their willingness to buy shiitake mushrooms, chicken eggs, and tuna fish from regions near the FDNPP through Figs. 5.33 to 5.36.

Figure 5.33 illustrates how the consumers' perceptions of the risk of radioactively contaminated food sold in grocery stores have impacts on the consumers'

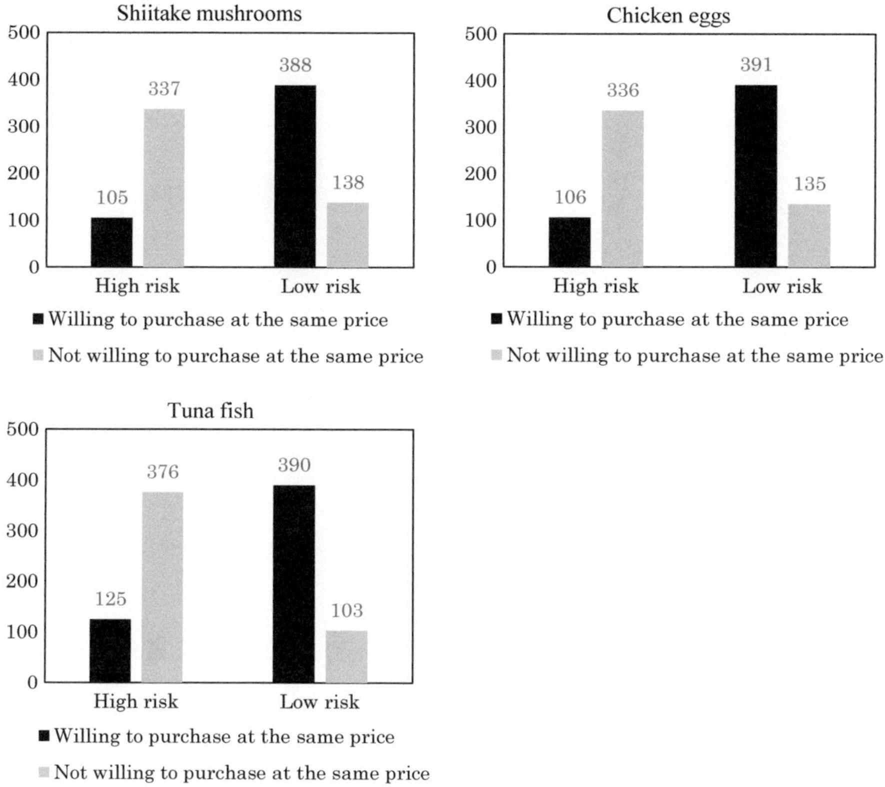

Fig. 5.33 Risk perceptions of food being contaminated with radioactive material and the willingness to buy II

willingness to buy shiitake mushrooms, chicken eggs, and tuna fish from regions near the FDNPP.

It is apparent from the figure that respondents who recognize highly of the risk regarding the radioactively contaminated food sold in grocery stores are more likely to react negatively about buying shiitake mushrooms, chicken eggs, and tuna fish from regions near the FDNPP. By contrast, most of the respondents who perceive such risk of food in grocery stores as low are reacting positively toward foods produced in regions near the FDNPP.

Hence, it can be implied that the difference in the perception of the risk of radioactively contaminated food sold in grocery stores does have an influence on the consumers' willingness to buy reactions.

Respondents' knowledge of radiation and their willingness to buy

Next in Fig. 5.34, we will investigate the relationship between the levels of possessions of respondents' knowledge of radiation and the respondents'

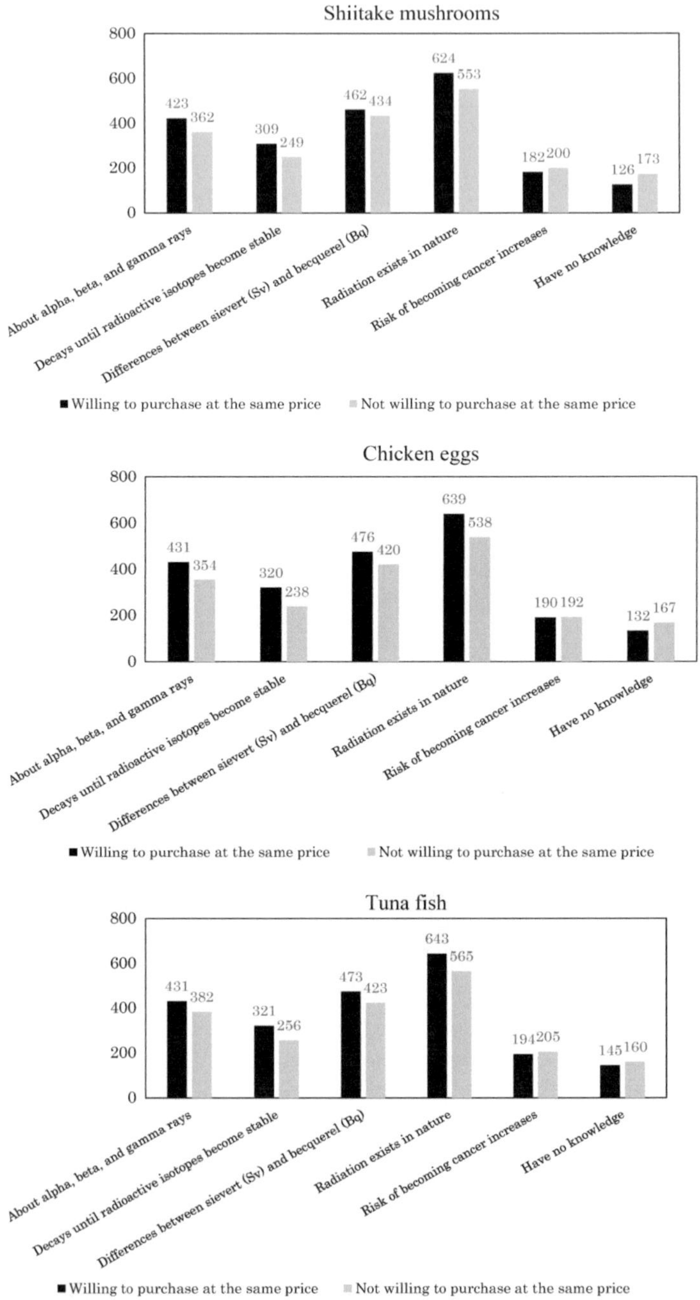

Fig. 5.34 Knowledge of radiation and the willingness to buy II

willingness to buy shiitake mushrooms, chicken eggs, and tuna fish from regions near the FDNPP.

You can see from the figure that respondents who have knowledge of radiation other than the "risk of becoming cancer increases" when additional amount of radiation received exceeds 100 mSv, more of the respondent are willing to buy food from regions near the FDNPP. It is apparent from the figure that more of the respondents among those with knowledge "about alpha, beta, and gamma rays," "radioactive isotopes continue to decay until they become stable isotopes," "differences between sievert and becquerel," and "radiation exists in nature" responded positively about buying food from regions near the FDNPP.

However, more respondents have a negative reaction toward food from the regions when they know that the "risk of becoming cancer increases" when an additional amount of radiation received exceeds 100 mSv. Respondents who picked "Have no knowledge" in radiation also have a negative response about buying food from the regions.

All in all, as mentioned in the section for cucumbers, apples, beef, and pork, here too it is important for the government to provide more opportunities like organizing education programs and seminars for the consumers to obtain knowledge of radiation and radioactive materials.

Respondents' perceptions of product origins and the willingness to buy

Figure 5.35 describes the relationship between the product origins that the respondents are concerned about the risk of radioactive contamination and the respondents' willingness to buy shiitake mushrooms, chicken eggs, and tuna fish from regions near the FDNPP.

From the figure, you can see that there is a discrepancy in the respondents' willingness to buy between those who have concerns about the product origin and those that do not care about the product origin of foods. Respondents who have concerns about the product origin have an adverse reaction about buying shiitake mushrooms, chicken eggs, and tuna fish from regions near the FDNPP while those that do not care about the product origin have a positive reaction to these foods.

Similarly to the foods of the previous subsection, it is obvious that more consumers are worried about the risk of radioactive contamination of food from Fukushima Prefecture and that they are more likely to avoid buying food from this prefecture.

Respondents' trust toward food safety standards for radiation and the willingness to buy

In this final subsection, I will discuss the relationship between the respondents' perceptions of radioactive contamination and their willingness to buy shiitake mushrooms, chicken eggs, and tuna fish from regions near the FDNPP. I will also explain how respondents' levels of trust on national food safety standards for radiation are related to the respondents' willingness to buy.

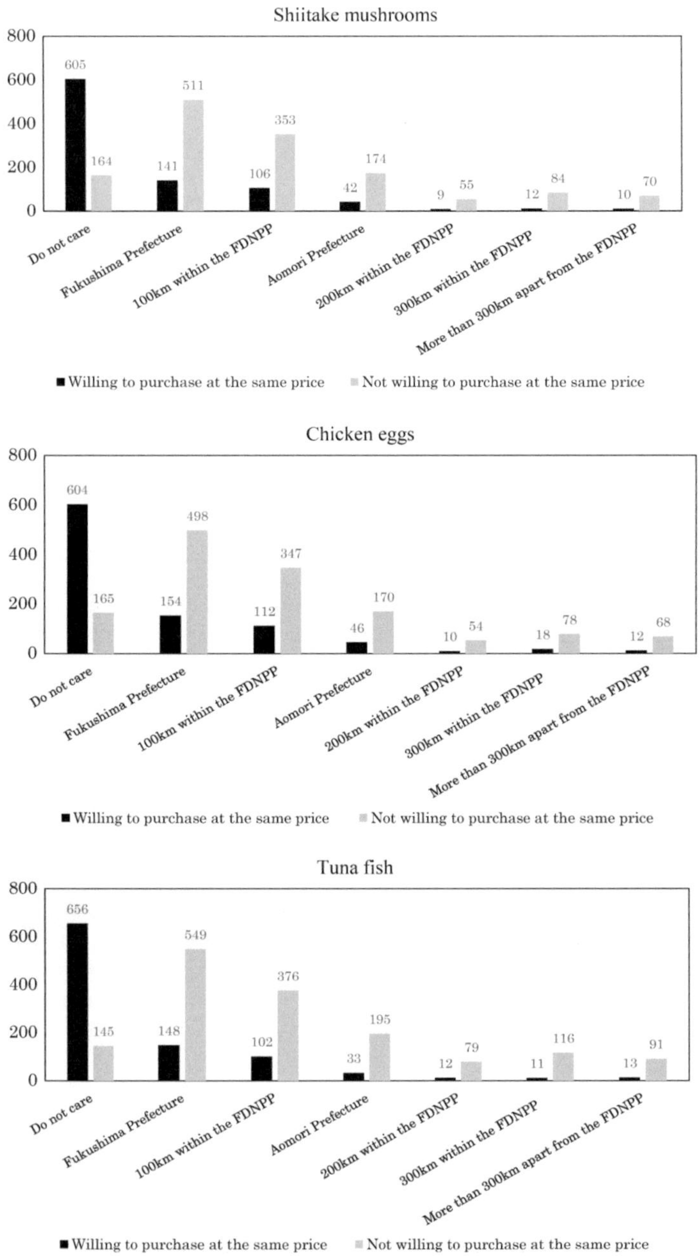

Fig. 5.35 Perceptions of product origin and the willingness to buy II

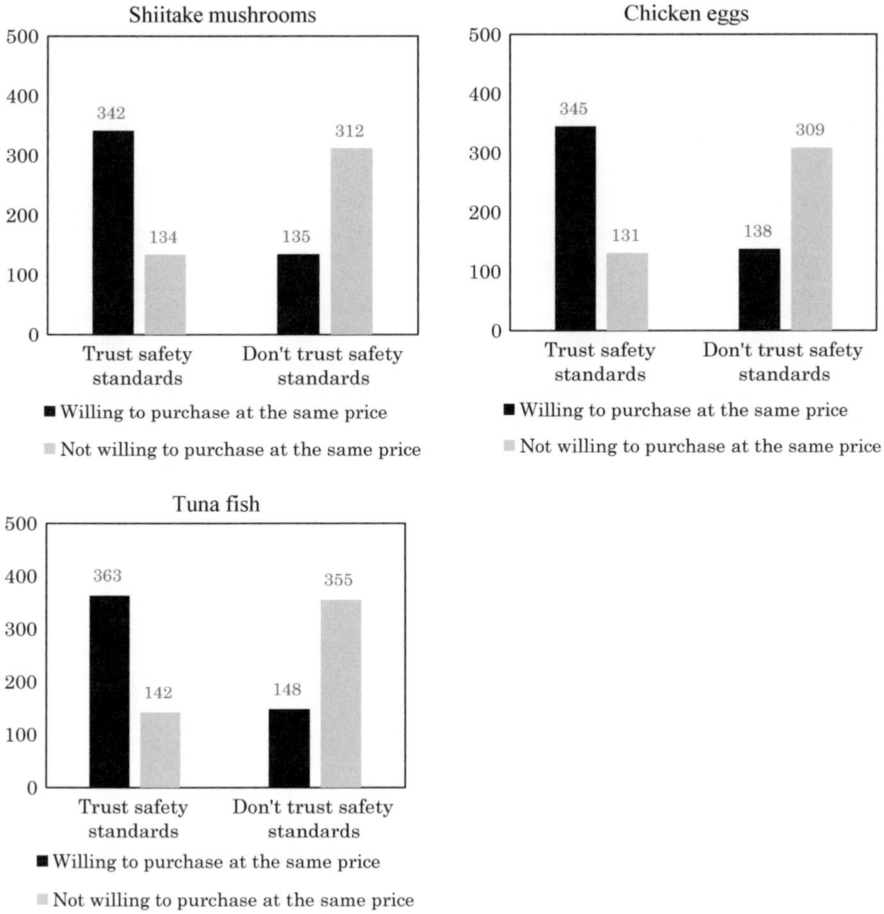

Fig. 5.36 Trust toward food safety standards for radiation and the willingness to buy II

As seen in Fig. 5.36, respondents who have high levels of trust in the national food safety standards for radiation have a positive response about buying food from regions near the FDNPP but those who do not trust the safety standards show a negative response about buying the same food.

Hence, the outputs presented in Fig. 5.36 imply that policies to improve the reliability of the food safety standards for radiation are necessary to enhance the sales of shiitake mushrooms, chicken eggs, and tuna fish from regions near the FDNPP.

5.3.5 Respondents' Perceptions of Willingness to Accept Food from Regions Near the FDNPP and Their Willingness to Buy

Fifth, I would like to investigate the respondents' willingness to accept perceptions for shiitake mushrooms, chicken eggs, and tuna fish from regions near the FDNPP and see how the respondents' levels of willingness to buy differ compared with other ten food products examined in this book.

Respondents' willingness to accept perceptions for shiitake mushrooms, chicken eggs, and tuna fish

Table 5.9 shows the percentages of respondents who are willing to buy shiitake mushrooms, chicken eggs, and tuna fish from regions near the FDNPP at the same price, at a 10% to 60% price discount, and not willing to buy at a 60% price discount compared with those from regions farther from the FDNPP.

Shiitake mushrooms, chicken eggs, and tuna fish are food products that about a half of the respondents showed willingness to buy food produced near the FDNPP at the same price as those apart from the FDNPP. Table 5.9 tells us that about a half of the respondents have a positive willingness to buy. The percentage of respondents who answered they would be willing to buy food from regions near the FDNPP with a 10–60% price discount is about 20%, but you can see from the figure that nearly 30% of the respondents are reluctant to buy these foods even at a 60% price discount. In particular, more than 30% of the respondents have a tendency to

Table 5.9 Respondents' willingness to accept perceptions for shiitake mushrooms, chicken eggs, and tuna fish from regions near the FDNPP

	Number of samples	Percentages %
Shiitake mushrooms		
Willing to buy at the same price	848	49.6
Willing to buy when the discount rate is between 10 and 60%	323	18.9
Not willing to buy even with a 60% discount rate	539	31.5
Chicken eggs		
Willing to buy at the same price	864	50.5
Willing to buy when the discount rate is between 10 and 60%	337	19.7
Not willing to buy even with a 60% discount rate	509	29.8
Tuna fish		
Willing to buy at the same price	893	50.5
Willing to buy when the discount rate is between 10 and 60%	358	20.2
Not willing to buy even with a 60% discount rate	517	29.2

avoid buying shiitake mushrooms even when there is a 60% price discount for the shiitake mushrooms from regions near the FDNPP.

As explained in Sect. 2.3, fungi such as the shiitake mushrooms are known to absorb radioactive substances from their surface cells and that they are prone to radioactive contamination. Therefore, the reason for having a higher avoidance rate for shiitake mushrooms among the survey respondents is perhaps because the respondents happened to learn such information that the risk of radioactive contamination for mushrooms are higher than other food products and it is likely that this information influenced their reactions toward shiitake mushrooms from regions near the FDNPP.

The effects of having safety labels on food products

Next, let's consider in Fig. 5.37 which indicates how many of the respondents are reluctant to buy shiitake mushrooms, chicken eggs, and tuna fish from regions

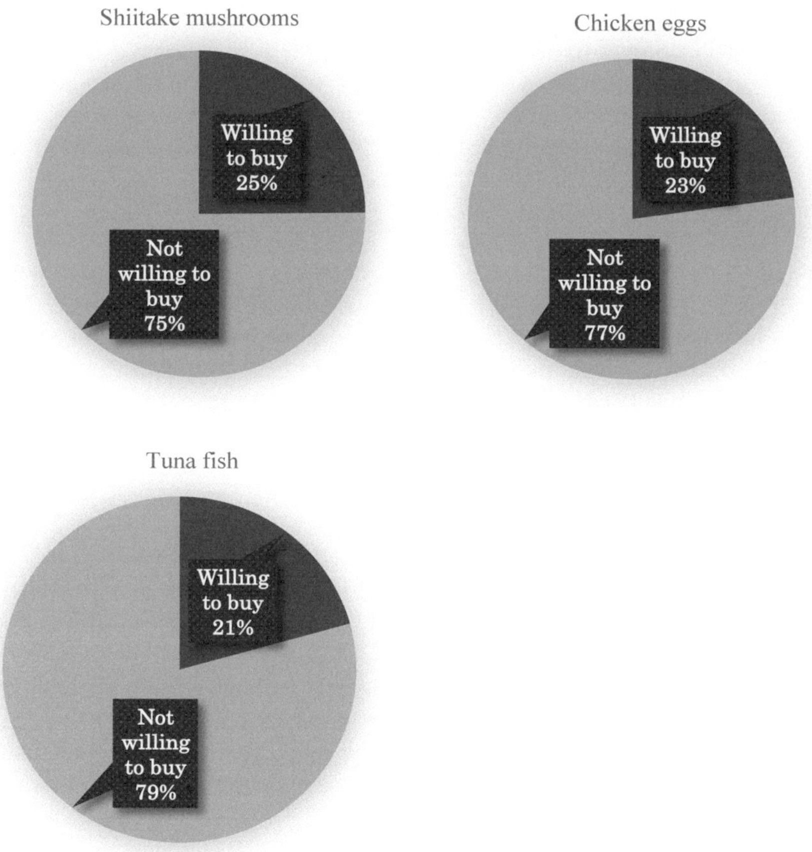

Fig. 5.37 Food labels that shows the product is safe from radiation and the willingness to buy II

near the FDNPP even when there is a 60% price discount. As shown in Table 5.9, this same group of respondents may change their answers if there is a label on the product showing that the product has no risk of radioactive contamination.

You can see from Fig. 5.37 that 75–79% of the respondents seem to continue avoiding food from regions near the FDNPP even there was a safety labels placed on the food products. By contrast, for shiitake mushrooms, one out of four respondents changed their answers and showed a positive response about buying shiitake mushrooms from regions near the FDNPP. Thus, it is believable that there are at least some chances to improve the consumers' willingness to buy shiitake mushrooms from regions near the FDNPP through creating and putting trustworthy safety labels on the product.

5.3.6 Respondents' Social Attributes and Their Willingness to Buy

Sixth, we will go over the relationship between the respondents' social attributes such as their resident locations and demographic attributes, and how such attributes have effects on the respondents' willingness to buy shiitake mushrooms, chicken eggs, and tuna fish from regions near the FDNPP. We will see the results presented in Fig. 5.38 through Fig. 5.45 for this purpose.

Respondents' residential distance from the FDNPP and the willingness to buy

Figure 5.38 exhibits the distance of the respondents' resident location from the FDNPP and how this is related to the respondents' willingness to buy food from regions near the FDNPP. Here too, the reason that we have more respondents who live 200–300 km apart and more than 400 km apart from the FDNPP is that most of the people in Japan reside in the three large metropolitan areas, Tokyo, Nagoya, and Osaka.

It is conceivable from the figure that respondents who live within 100 km apart from the FDNPP have a positive reaction about buying food from regions near the FDNPP but among those who live more than 100 km apart from the FDNPP only about a half of the respondents show a positive reaction. Comparing this results with Fig. 5.17 of the previous section, it is likely that respondents are reacting more negatively about buying shiitake mushrooms, chicken eggs, and tuna fish compared with cucumbers, apples, beef, and pork.

It is presumable that this gap in the consumers' willingness to buy the foods we deal with in this section as opposed to in the previous section, is perhaps because shiitake mushrooms, chicken eggs, and tuna fish are produced throughout the country. Due to this, it is not necessary for consumers to deliberately buy these products from regions near the FDNPP.

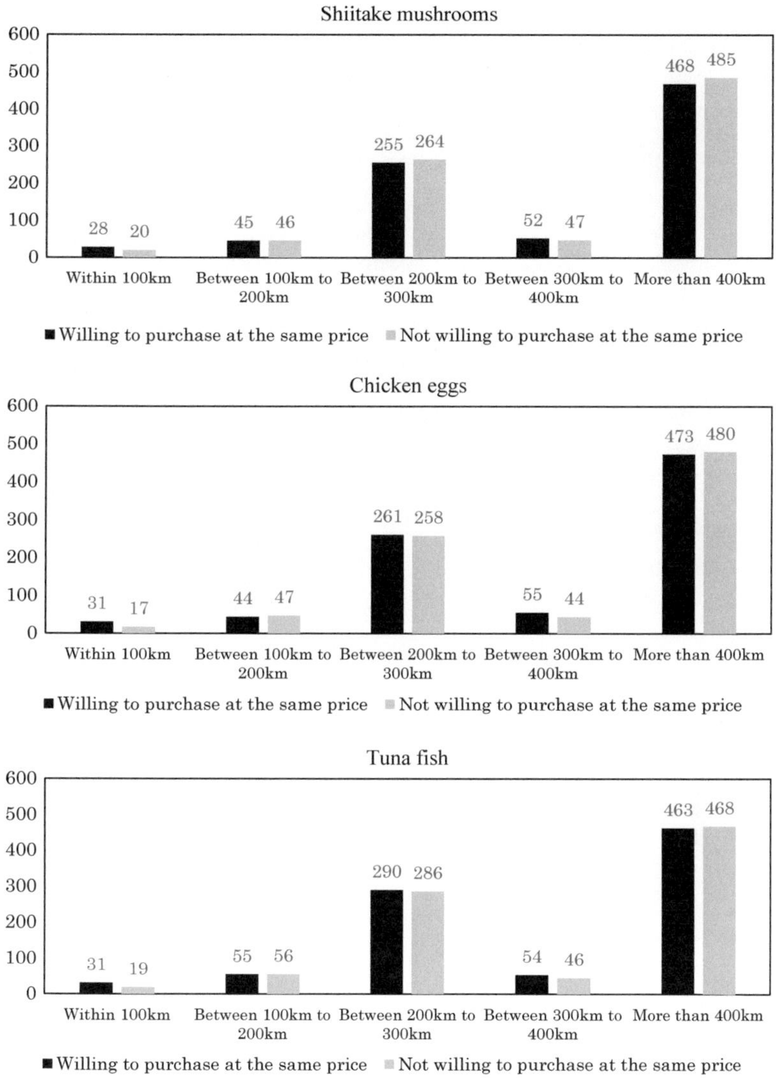

Fig. 5.38 Location of residence and the willingness to buy II

Respondents' gender and their willingness to buy

Figure 5.39 shows how the respondents' willingness to buy shiitake mushrooms, chicken eggs, and tuna fish from regions near the FDNPP is different between their genders.

It is clear from the figure that female respondents are more likely to react negatively about buying food from these regions compared with the male respondents. The reason for this difference in the reaction between the female and male respondents is probably because, as explained in the previous subsection, female respondents spend more time with their children than male respondents in Japan and that they are more concerned about the risk of their children being exposed to radioactive materials.

Respondents' age and their willingness to buy

Next in Fig. 5.40, I would like to describe the relationship between the respondents' age and their willingness to buy shiitake mushrooms, chicken eggs, and tuna fish from regions near the FDNPP.

It is apparent from the figure that more respondents who are below age 40 are not willing to buy food from regions near the FDNPP. Like we saw in Fig. 5.19 for cucumbers, apples, beef, and pork, the percentage of respondents of age 30s

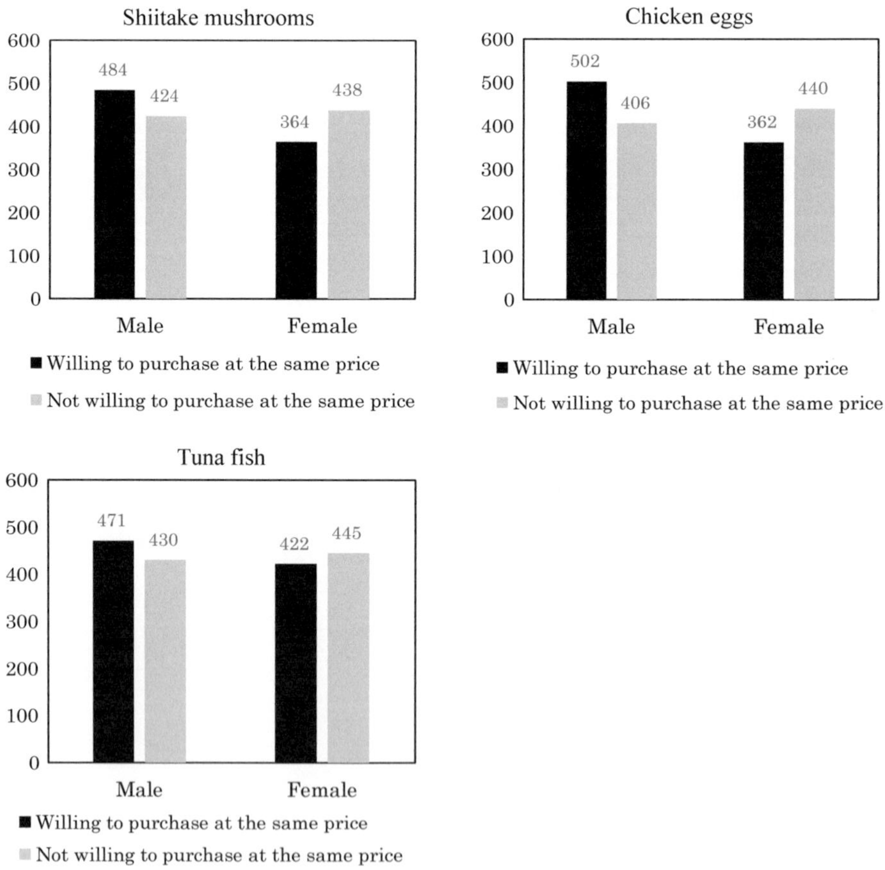

Fig. 5.39 Gender and the willingness to buy II

reacting negatively about buying shiitake mushrooms, chicken eggs, and tuna fish from regions near the FDNPP is higher compared with respondents of other age categories. The reason for this is perhaps because age 30s is the age group having little children and that individuals of this age group are concerned about their children getting exposed to radiation.

Furthermore, you can see from the figure that more of the respondents who are in their 50s and 60s are reacting positively about buying food from regions near the FDNPP.

Hence, it is conceivable that as respondents' age increases the likelihood of respondents accepting food from these regions increases, which implies that age has a positive effect regarding the consumers' willingness to buy food from these regions.

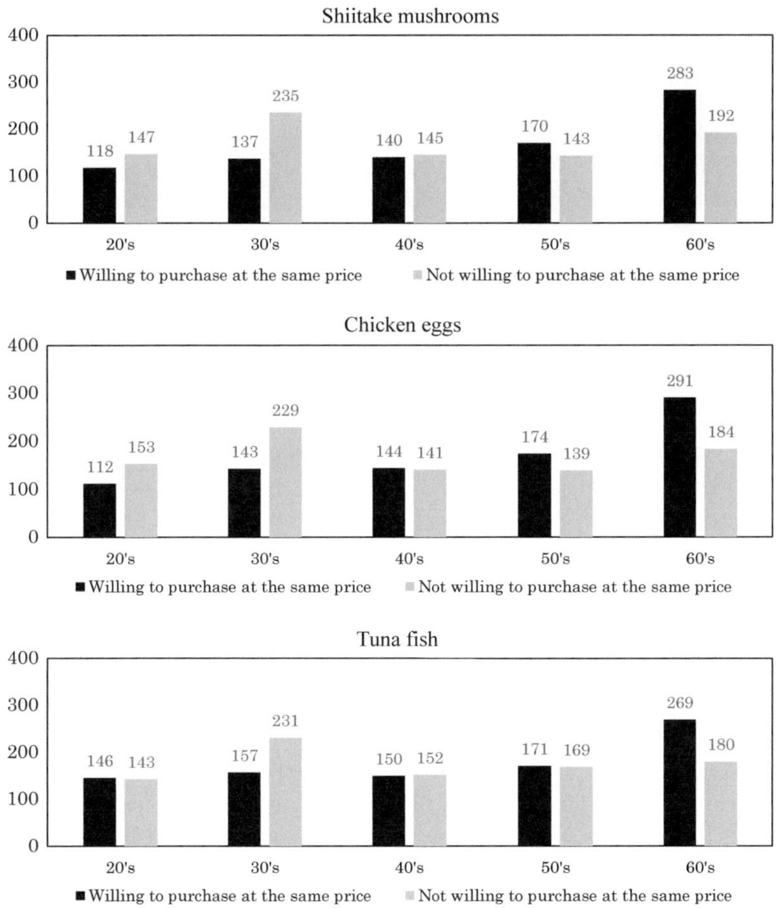

Fig. 5.40 Age and the willingness to buy II

Relationships between the presence or absence of children and the willingness to buy

Figures 5.41 and 5.42 illustrate the relationship between respondents with children in their households and their willingness to buy shiitake mushrooms, chicken eggs, and tuna fish from regions near the FDNPP.

Figure 5.41 describes how the respondents' structure of family members has an influence on the respondents' willingness to buy food from these regions. From the figure, you can see that respondents who live with children under the age of elementary school have a higher number of respondents reacting negatively about buying food from these regions. In particular, as seen in the figures for shiitake mushrooms and chicken eggs, it is believable that the consumers were particularly worried about the risk of effects from radiation on little children because respondents with high school age children are more likely to buy food from regions near the FDNPP. However, for tuna fish, respondents with high school children are also reacting negatively toward food from these regions suggesting that consumers are evaluating highly of the radioactive contamination risk involved with tuna fish.

In Fig. 5.42, we will go over the relationship between the number of children under age 15 in respondents' household and the respondents' willingness to buy shiitake mushrooms, chicken eggs, and tuna fish from regions near the FDNPP. Similarly with food products we discussed in the previous subsection, respondents having more than two children that are under age 15 are more likely to react negatively about buying food from these regions.

From the above investigations, consumers with little children are more likely to avoid buying food from regions near the FDNPP and it is presumable that such consumers consider seriously about the radioactive contamination risk of food produced in these regions.

Educational level and the willingness to buy

Figure 5.43 presents how the respondents' educational levels have impacts on the respondents' reaction toward shiitake mushrooms, chicken eggs, and tuna fish from regions near the FDNPP. It is evident from the figure that compared with the respondents of other education groups, those with a bachelor's degree have a high tendency to avoid buying food from these regions. This indicates that respondents' educational levels have an adverse impact on the respondents' willingness to buy food from these regions.

I believe the reason why the respondents with higher educational levels react negatively about buying food from regions near the FDNPP is related to the difference in the knowledge of radiation they hold between the education groups. To confirm this assumption, I checked the percentages of respondents who know that the risk of becoming cancer increases when individual's additional amount of radiation received exceeds 100 mSv among the different education groups. The results of this tells us that compared to respondents of other education groups, those with a bachelor's degree have a higher percentage of individuals who know that the risk of

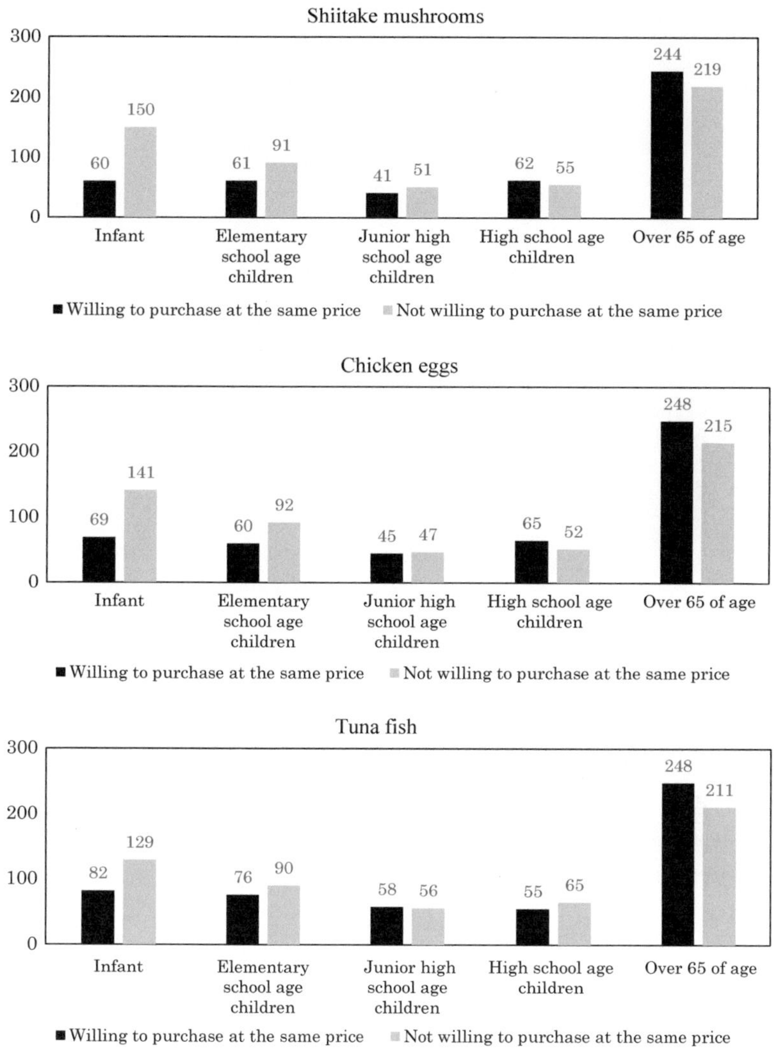

Fig. 5.41 Family structure and the willingness to buy II

becoming cancer increases when exposed to radiation above the 100 mSv level.[3] Hence, it is likely that consumers with a higher educational level are inclined to avoid buying shiitake mushrooms, chicken eggs, and tuna fish from regions near the

[3]For example, among the respondents who answered the survey for shiitake mushrooms and chicken eggs and have high school, junior college, and bachelor's degree educational levels, 19.5, 20.3, and 24.3% have knowledge regarding the risk of becoming cancer when exposed to radiation.

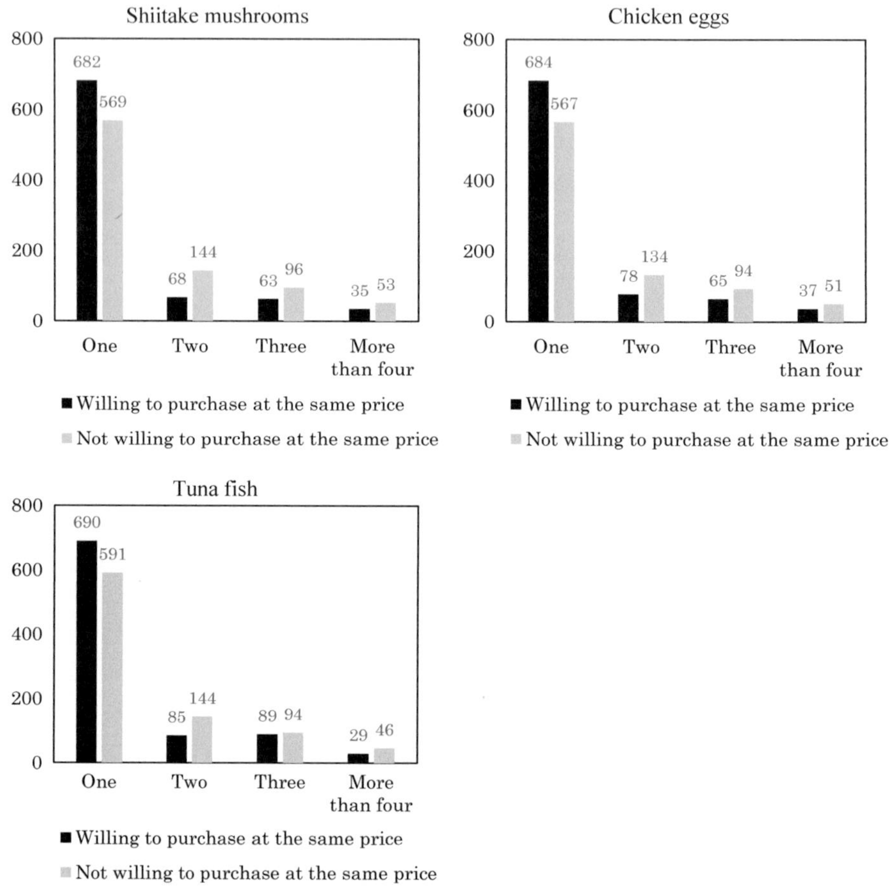

Fig. 5.42 Number of children under age 15 and the willingness to buy II

FDNPP because they are more knowledgeable about the risk of radiation compared with those with a lower educational level.

Respondents' income level and the willingness to buy

As the final part of this subsection, I would like to examine how respondents' income levels have an influence on the respondents' willingness to buy shiitake mushrooms, chicken eggs, and tuna fish from regions near the FDNPP. We will use Fig. 5.44 for this purpose.

First, for the respondents with an annual income of below 2 million yen, more of them refused to buy food from regions near the FDNPP. This is likely related to having more female respondents in this income tier as explained in the previous subsection.

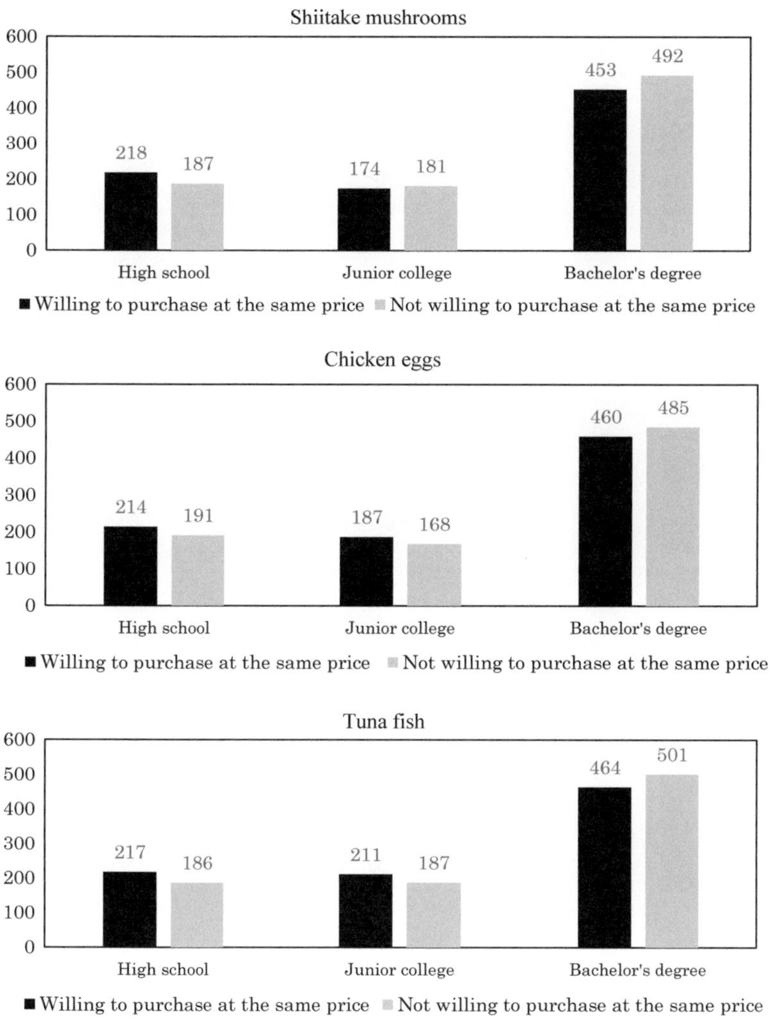

Fig. 5.43 Highest educational level and the willingness to buy II

Second, there is a difference in the willingness to buy the food products when the respondents' income levels are between 2 and 4 million yen. While more of the respondents are willing to buy eggs and tuna fish from regions near the FDNPP, comparatively fewer respondents reacted positively about purchasing shiitake mushrooms. As seen in Table 5.9, this difference is perhaps attributed to more respondents reacting negatively toward shiitake mushrooms produced in regions near the FDNPP.

Third, among the respondents with an annual income of 4–8 million yen, more of the respondents show negative response about buying food from regions near the

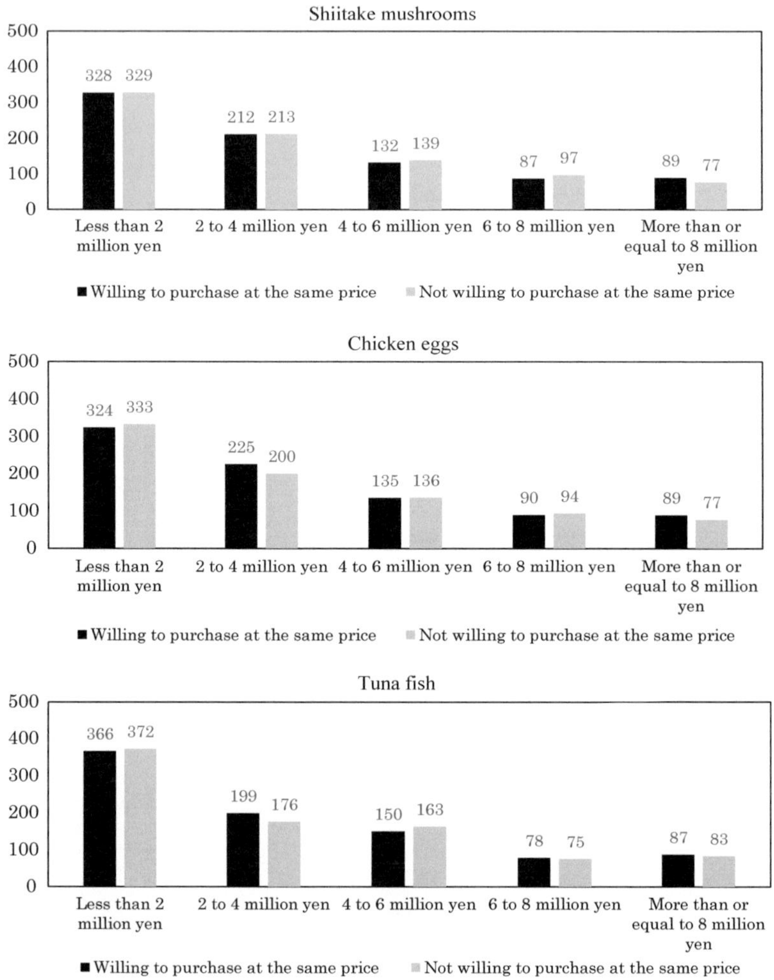

Fig. 5.44 Income level and the willingness to buy II

FDNPP. I believe the reason for this is because respondents who belong to this income tier are mostly in their 30s and 40s and such respondents are more likely to have little children in their homes.

Finally, for respondents with an annual income of more than or equal to 8 million yen, it can be seen from the figure that more of the respondents have positive reaction toward food from regions near the FDNPP. It is conceivable that this is because 80% of the survey respondents that belong to this income tier are male and more than half of them are older than 50 years of age.

5.4 Food and Beverage Produced in Regions Near the FDNPP that a Majority of the Respondents Are not Willing to Buy: Rice, Mineral Water, and Wakame Seaweed

In this subsection, as with previous ones, we will examine how the following six factors have an influence on food and beverage produced in regions near the FDNPP that a majority of the respondents show a negative response toward the food and beverage products. The six factors are respondents' eating habits, perceptions of food safety, interests in social problems, attitudes toward nuclear contamination, consumers' willingness to accept products produced near the FDNPP, and social attributes.

Rice, mineral water, and wakame seaweed are food and beverage we investigate among the ten products of our interest in this subsection. These three products have the lowest willingness to buy among the ten products, that more than a half of the respondents are not willing to buy food and beverage from regions near the FDNPP at the same price as those produced in regions apart from the FDNPP.

5.4.1 Respondents' Eating Habits and Their Willingness to Buy

First, let's discuss in Figs. 5.45 and 5.46 the relationship between the respondents' eating habits and their willingness to buy rice, mineral water, and wakame seaweed from region regions near the FDNPP.

Shopping frequency and willingness to buy

Correspondingly, as with previous subsections, Fig. 5.45 illustrates the respondents' shopping frequency and its relationship with the willingness to buy food and beverage from regions near the FDNPP.

According to the figure for rice, respondents who go grocery shopping almost every day and a few days a week tend to respond negatively about buying rice from regions near the FDNPP while those who only go grocery shopping a few days a month have a positive reaction toward rice from these regions. For mineral water and wakame seaweed, a similar relationship is apparent between the respondents' frequency for grocery shopping and their willingness to buy food and beverage from regions near the FDNPP. More of the respondents who often go grocery shopping are not willing to buy mineral water and wakame compared with those who rarely go grocery shopping.

Hence, it is likely that the consumers' shopping frequency has a negative effect on their willingness to buy reactions toward rice, mineral water, and wakame seaweed from region regions near the FDNPP.

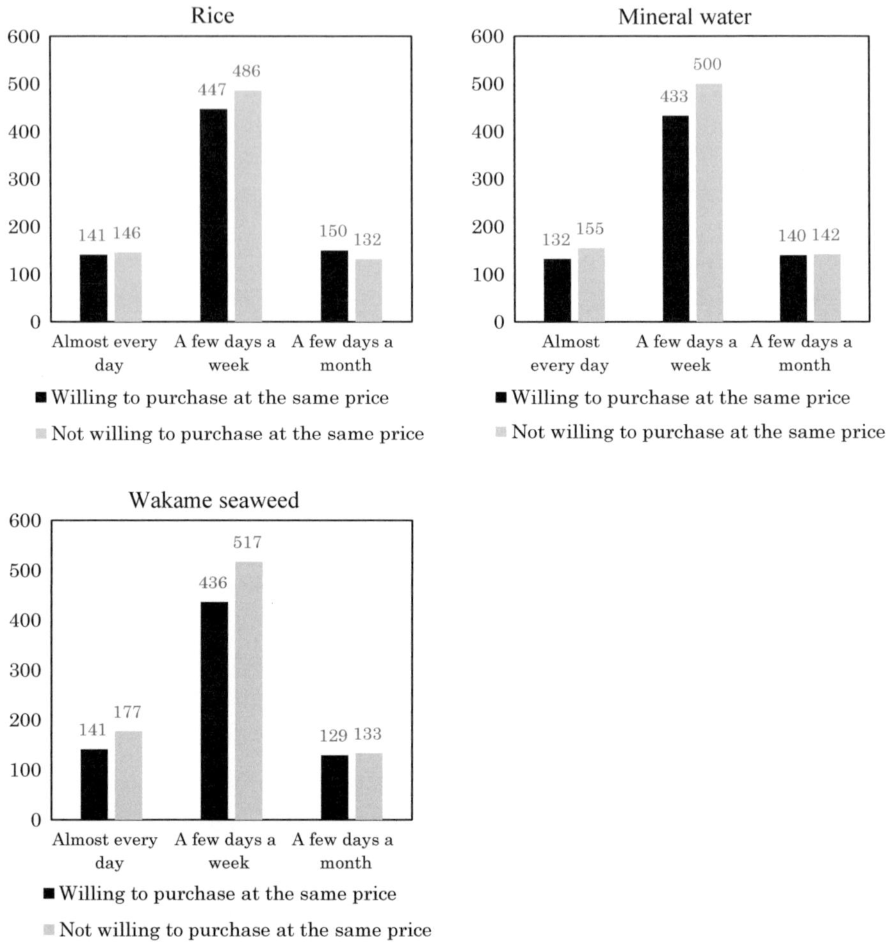

Fig. 5.45 Shopping frequency and the willingness to buy III

Purchase experience and the willingness to buy

Figure 5.46 depicts the relationship between the respondents' purchase experiences of rice, mineral water, and wakame seaweed and the respondents' willingness to buy these foods and beverages from regions near the FDNPP. As seen in the figure, survey respondents with purchase experiences of rice and mineral water have the tendency to respond negatively if these products are from regions near the FDNPP. For wakame seaweed, more respondents are not willing to buy wakame seaweed from regions near the FDNPP even among the respondents that do not buy

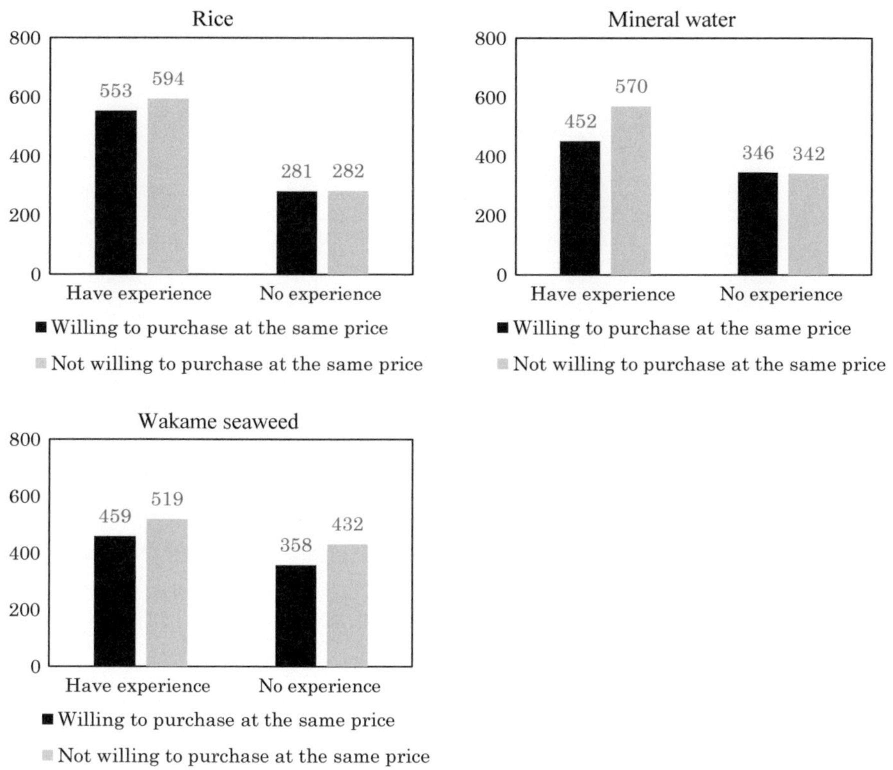

Fig. 5.46 Purchase experience and the willingness to buy III

wakame seaweed for themselves. This indicates that the consumers' are highly concerned about the risk of radioactive contamination for wakame seaweed produced in regions near the FDNPP.

Cooking frequency and the willingness to buy

Figure 5.47 presents how the difference in respondents' cooking frequency has impacts on their willingness to buy rice, mineral water, and wakame seaweed from regions near the FDNPP. From the figure, it is clear that more respondents who cook almost every day and a few days a week are not willing to buy food and beverage from regions near the FDNPP at the same price as that from regions apart from the FDNPP. However, compared to respondents with a high cooking frequency (cook almost everyday and a few days a week), those with a low cooking frequency (cook a few days a month and not so often) are more likely to respond positively about buying food and beverage from those regions.

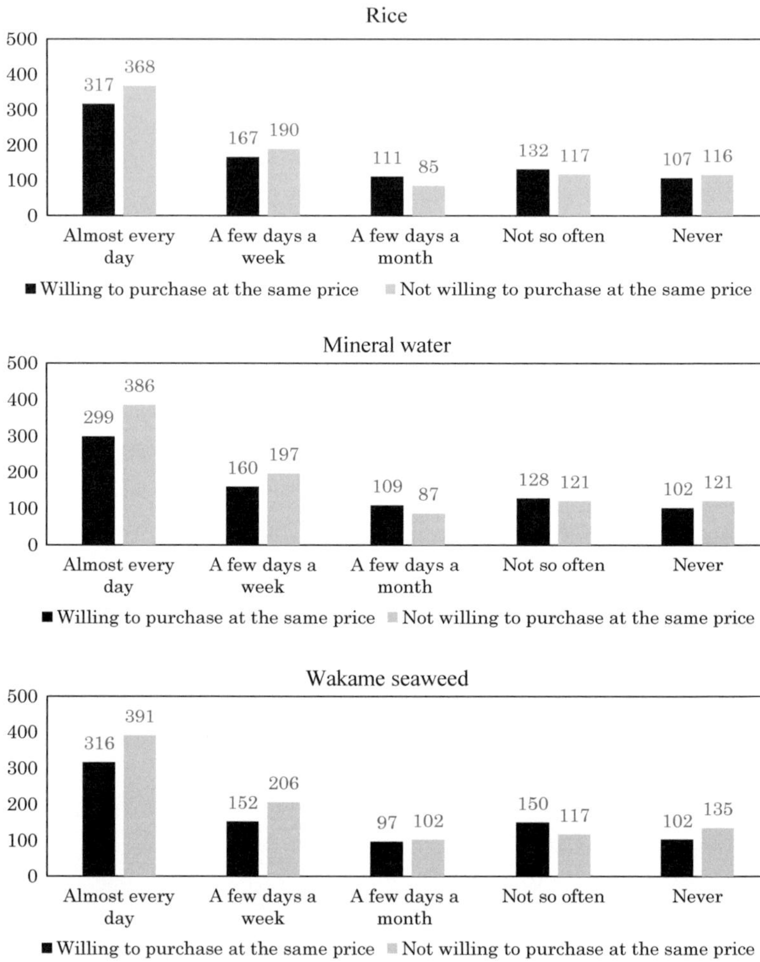

Fig. 5.47 Cooking frequency and the willingness to buy III

Therefore, the figure indicates that the cooking frequency of consumers has a negative influence on the their willingness to buy rice, mineral water, and wakame seaweed from regions near the FDNPP.

Eating style and the willingness to buy

Figure 5.48 describes how respondents' difference in their eating styles has impacts on their willingness to buy rice, mineral water, and wakame seaweed from regions near the FDNPP.

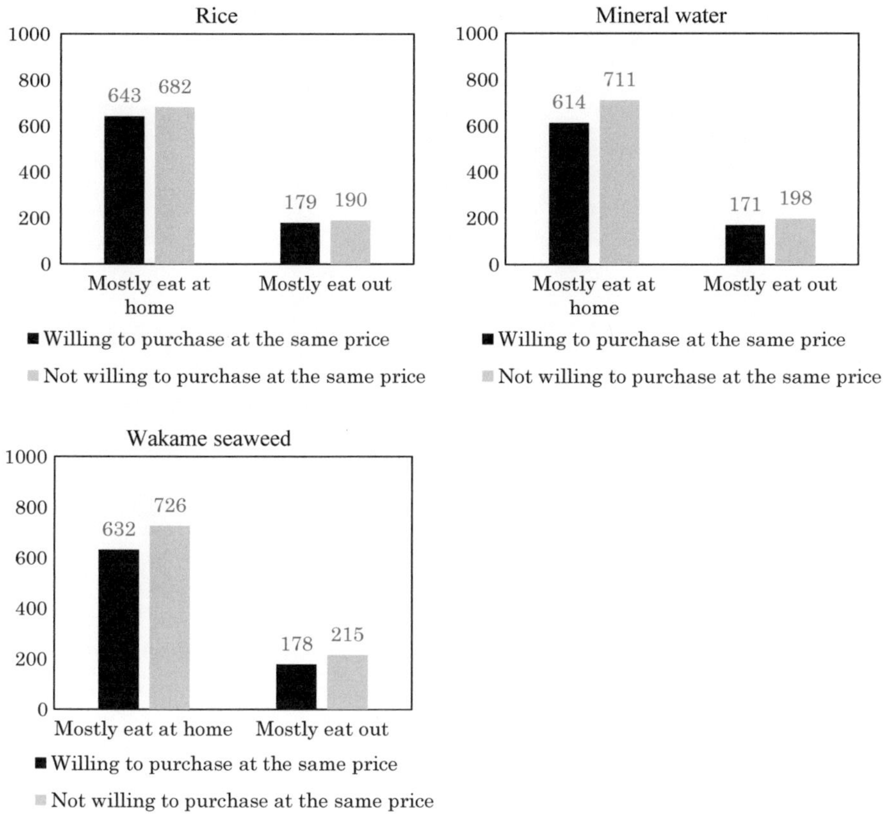

Fig. 5.48 Eating style and the willingness to buy III

In figures of all products, whether the respondents normally eat at home or outside, more respondents are showing a negative willingness to buy the products. Thus, it is conceivable that the difference in the respondents' eating style does not have an influence on their willingness to buy rice, mineral water, and wakame seaweed from regions near the FDNPP.

Respondents' concerns when buying food and the willingness to buy

Figure 5.49 shows the relationship between the factors that the respondents are concerned about when purchasing food products and the respondents' willingness to buy rice, mineral water, and wakame seaweed from regions near the FDNPP.

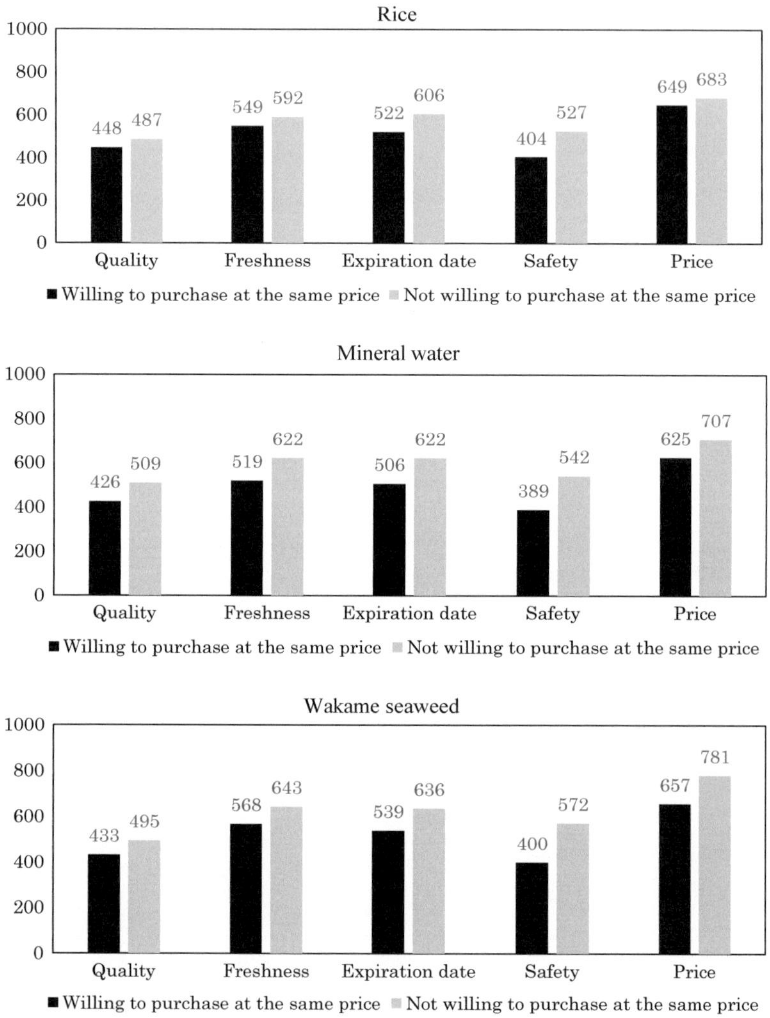

Fig. 5.49 Factors that consumers worry about when buying food and the willingness to buy III

From the figure, it is evident that for all products respondents are reacting negatively toward food and beverage produced in regions near the FDNPP. In particular, respondents who are worried about the safety issue when purchasing food products have a higher avoidance rate toward these products compared with those who selected other factors.

5.4.2 Respondents' Perceptions of Food Safety Issue and Their Willingness to Buy

Second, I would like to examine how differences in the respondents' perceptions of the food safety issue affect their willingness to buy rice, mineral water, and wakame seaweed from regions near the FDNPP. We will go over Figs. 5.50 and 5.51 for this objective.

Respondents' perceptions of unsafe food and the willingness to buy

Figure 5.50 shows the relationship between the types of actions respondents take to keep themselves away from unsafe food and the respondents' willingness to buy food and beverage from regions near the FDNPP. According to the figure, respondents who take some actions have an adverse reaction toward products of regions near the FDNPP. It seems that more respondents check the product origin and the expiration and best buy dates when they buy a grocery, and here too, more respondents are not willing to buy products from regions near the FDNPP.

On the other hand, although the numbers are small, respondents who do nothing special to keep themselves away from unsafe food are more willing to buy products from regions near the FDNPP.

Hence, similarly with the food products, we examined in the previous subsections, it can be implied from Fig. 5.50 that consumers who are highly conscious about keeping themselves free from unsafe food are less willing to buy food and beverage from regions near the FDNPP.

Trust in food labels and the willingness to buy

Here, I would like to explain how differences in the respondents' confidence toward food labels are related to their willingness to buy rice, mineral water, and wakame seaweed from regions near the FDNPP.

For rice, it is clear from Fig. 5.51 that more respondents with trust in food labels are willing to buy rice from regions near the FDNPP. On the contrary, for mineral water and wakame seaweed, even among the respondents with trust in food labels are showing a negative response about buying these products if they are from regions near the FDNPP. Furthermore, respondents who do not trust food labels have a negative response toward all three products investigated in this subsection.

Therefore, for rice, there is a possibility to enhance the sales of rice produced in regions near the FDNPP by putting safety labels on the product to show that it is safe from radioactive contamination. However, it seems that the effects of placing such safety labels for mineral water and wakame seaweed are limited suggesting that transmitting safety information through labeling is not effective for restoring the trust on mineral water and wakame seaweed produced in regions near the FDNPP.

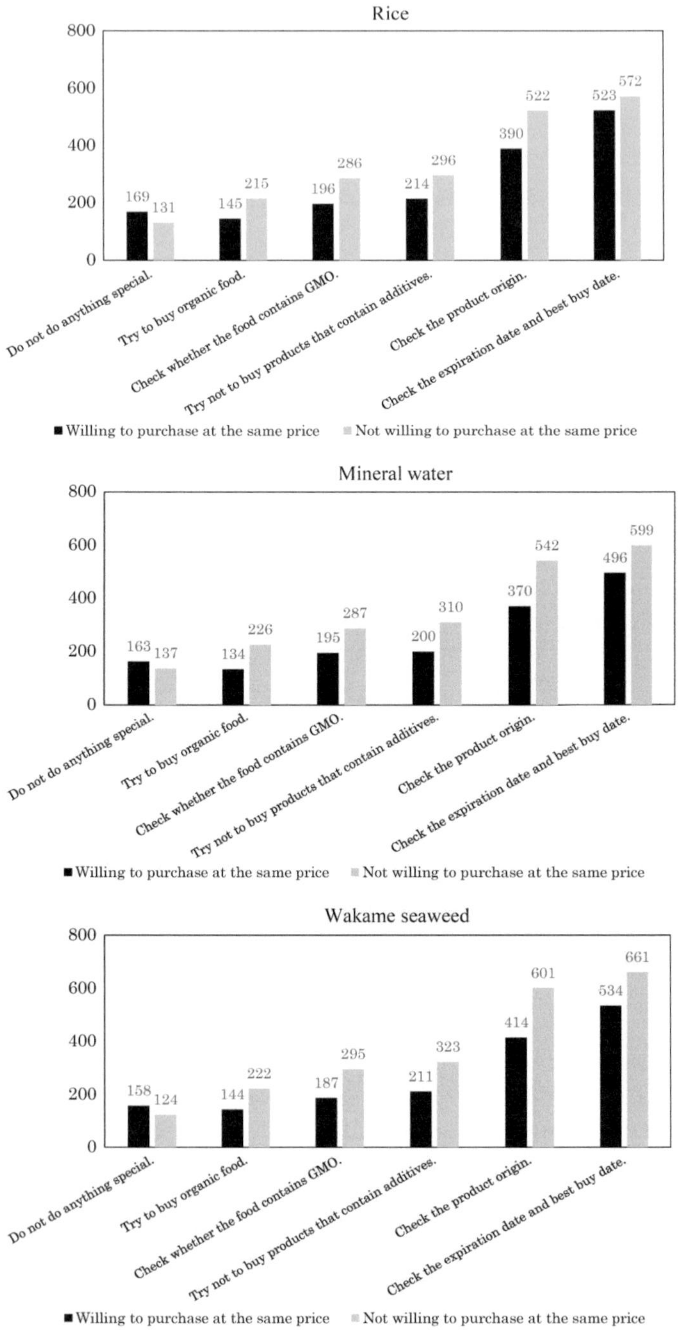

Fig. 5.50 Consumers' actions to keep themselves away from unsafe food and the willingness to buy III

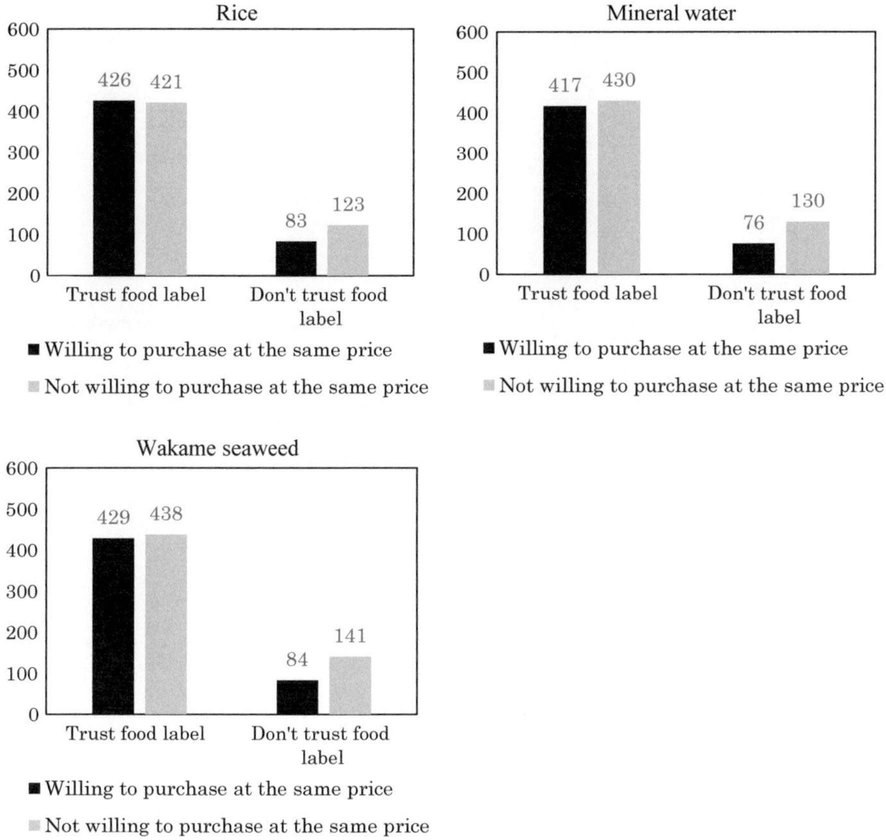

Fig. 5.51 Trust in food labels and the willingness to buy III

5.4.3 Respondents' Interests in Social Problems and the Willingness to Buy

Third, I would like to seek the relationship between the respondents' interests in social issues and their willingness to buy rice, mineral water, and wakame seaweed from regions near the FDNPP through Figs. 5.52 and 5.53.

First, Fig. 5.52 illustrates how the respondents' eagerness to help preserve the environment is related to their willingness to buy food and beverage products from regions near the FDNPP. From the figure, respondents that are more willing to attend environmental activities have a positive reaction toward food and beverage from the regions near the FDNPP. On the other hand, respondents who are not interested in preserving the environment have a negative reaction about buying rice, mineral water, and wakame seaweed from these regions.

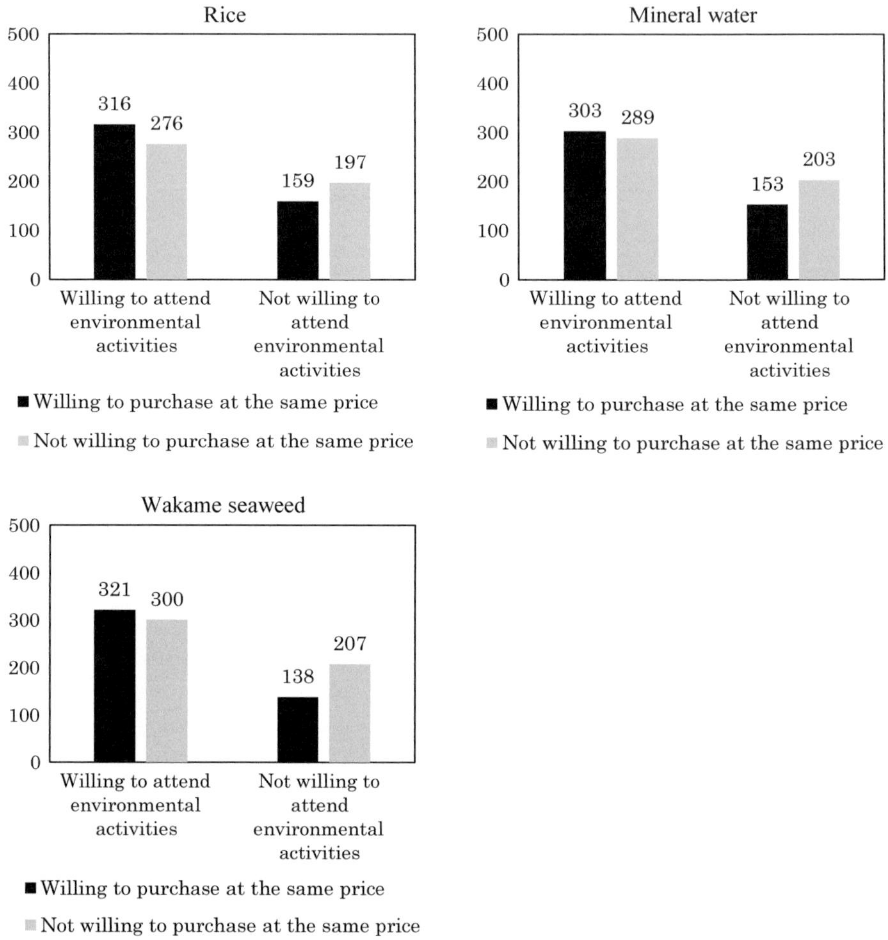

Fig. 5.52 Aggressiveness in preserving the environment and the willingness to buy III

Next, in Fig. 5.53, I would like to explain the relationship between the respondents' aggressiveness in reconstructing the regions suffered from the Tohoku-Pacific Ocean Earthquake and their willingness to buy products from the FDNPP. It is evident that in all products respondents that are ardent about helping the disaster regions to recover from their damage after the earthquake are more willing to buy food and beverage from regions near the FDNPP. On the other hand, those who are reluctant to help the disaster regions are less likely to buy food and beverage from these regions.

It is conceivable from the above results that consumers with more interests in social problems have a positive reaction about buying rice, mineral water, and wakame seaweed from regions near the FDNPP. As also explained in the previous subsection, I believe such consumers with high social interests are altruistic

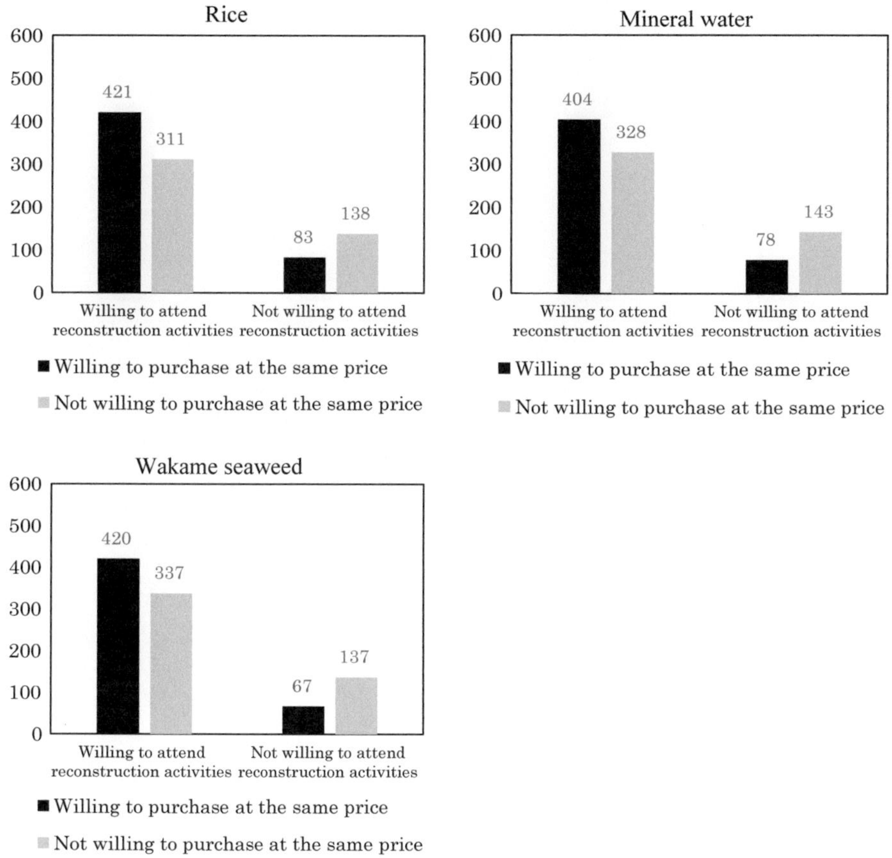

Fig. 5.53 Aggressiveness in reconstructing the regions suffered from the disaster and the willingness to buy III

consumers that are enthusiastic about helping other people. Therefore, performing an advertisement to target such altruistic consumers might help increase the sales of food and beverage products of regions near the FDNPP.

5.4.4 Respondents' Perceptions of Radioactive Contamination and Their Willingness to Buy

Fourth, I would like to describe how respondents' perceptions of radioactive contamination are related to the respondents' willingness to buy rice, mineral water, and wakame seaweed from regions near the FDNPP. We will go over Figs. 5.54 to 5.57 for this purpose.

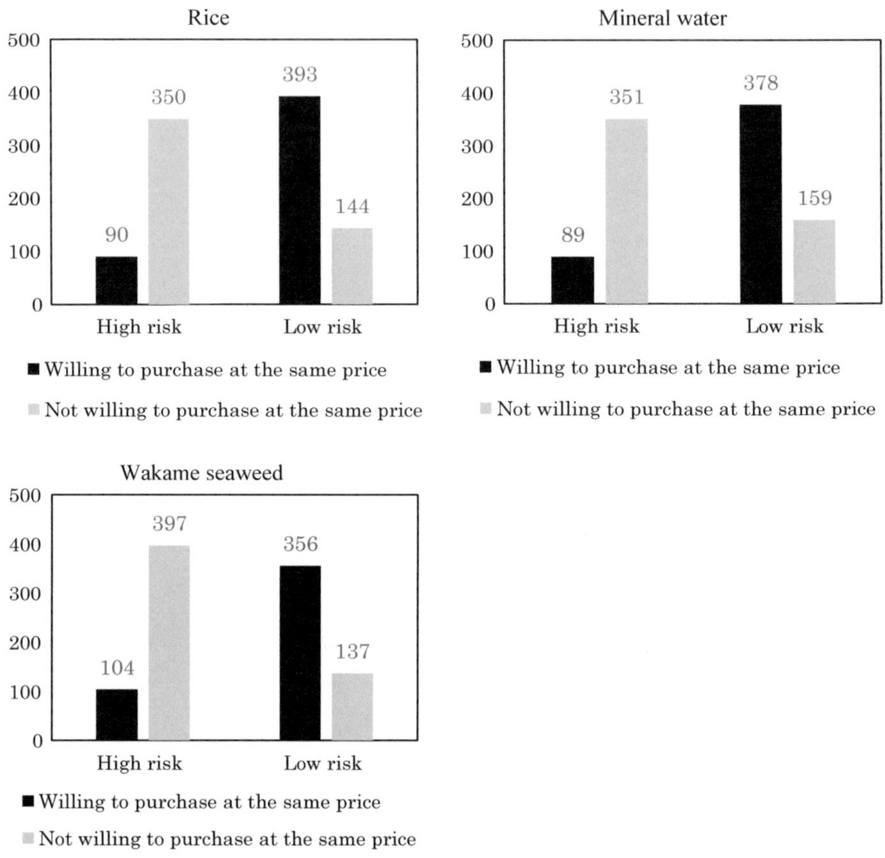

Fig. 5.54 Risk perceptions of food being contaminated with radioactive material and the willingness to buy III

Respondents' risk perceptions toward radioactively contaminated food and their willingness to buy

Figure 5.54 shows the relationship between the respondents' risk perceptions toward radioactively contaminated food and their willingness to buy rice, mineral water, and wakame seaweed from regions near the FDNPP.

You can see from the figure that respondents who think highly of the risk regarding the radioactive contamination of food sold at grocery stores are more likely to avoid buying food and beverage from regions near the FDNPP. By contrast, those who consider little of such risk involved in food sold at grocery stores have a positive reaction about buying food from these regions.

Therefore it is probable to think that consumers who have serious concerns about the risk of buying radioactively contaminated food at a grocery store have a lower willingness to buy rice, mineral water, and wakame seaweed from regions near the FDNPP.

Respondents' knowledge of radiation and their willingness to buy

Figure 5.55 presents how the respondents' difference in the type of radiation knowledge they hold is related to their willingness to buy rice, mineral water, and wakame seaweed from regions near the FDNPP.

From the figures of rice and mineral water, it is apparent that more of the respondents with type of knowledge "about alpha, beta, and gamma rays" and radiation "decays until radioactive isotopes become stable" have a positive reaction toward these products. In contrast, respondents having type of knowledge about "Differences between sievert (Sv) and becquerel (Bq)" and "Risk of becoming cancer increases" are more likely to respond negatively about buying rice and mineral water from these regions. Thus, it is probable that the respondents' willingness to buy rice and mineral water is different between the above types of radiation knowledge possessed by the consumers.

For wakame seaweed, besides the knowledge radiation "decays until radioactive isotopes become stable," more of the respondents with knowledge of radiation have a negative willingness to buy. This implies that a large number of respondents are concerned about the risk of radioactive contamination for wakame seaweed including those with radiation knowledge.

Finally, here too, for all products respondents with no knowledge have a high avoidance rate toward food and beverage from regions near the FDNPP. This indicates that it is important to create more opportunities for the consumers to learn about radiation and radioactive materials.

Respondents' perceptions of product origins and the willingness to buy

Figure 5.56 exhibits the relationship between the product origins that the respondents are concerned about for the risk of radioactive contamination, and the respondents' willingness to buy rice, mineral water, and wakame seaweed from regions near the FDNPP.

As seen in the figure, respondents selecting any of the choices regarding the product origins provided in the survey are more likely to avoid food and beverage from regions near the FDNPP. On the other hand, those that do not have concerns in the product origins show a positive willingness to buy products from these regions.

Hence, similarly, with the results of the previous subsections, it is presumable that whether or not the respondents show a positive willingness to buy food and beverage of regions near the FDNPP depends on their anxieties about the product origins.

Respondents' trust toward food safety standards for radiation and the willingness to buy

Figure 5.57 illustrates the relations between the respondents' levels of trust they put on the national food safety standards for radiation and their willingness to buy rice, mineral water, and wakame seaweed from regions near the FDNPP.

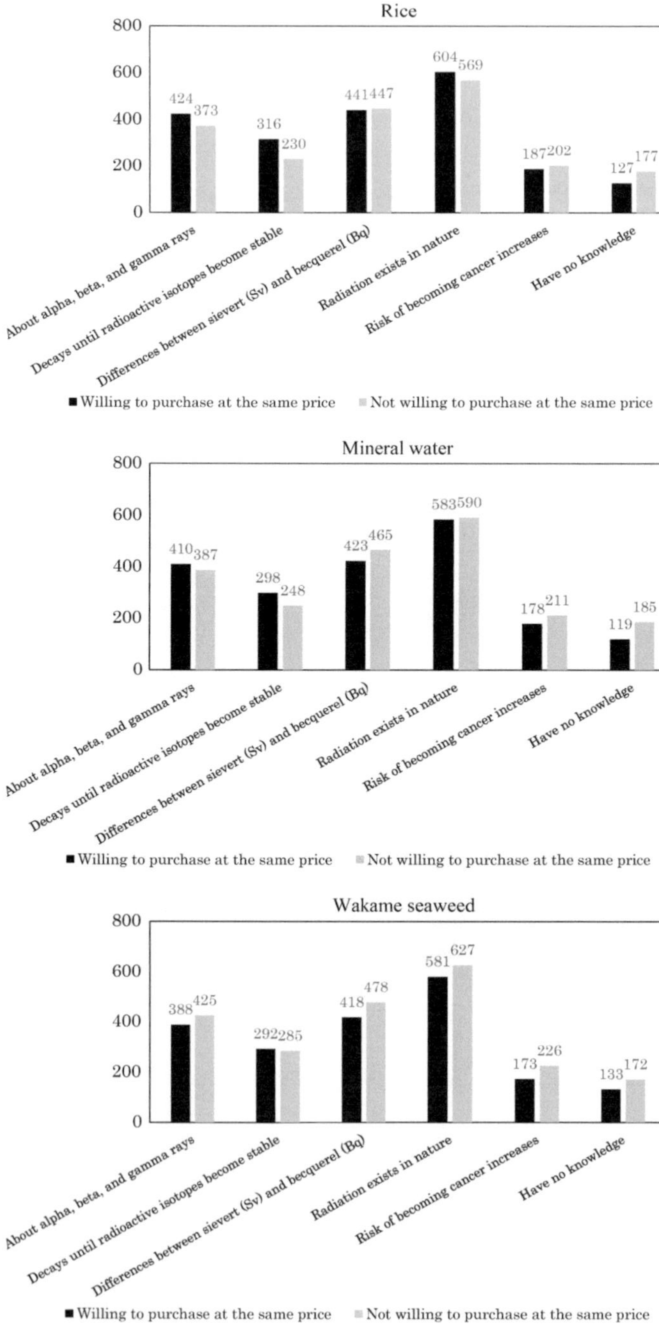

Fig. 5.55 Knowledge of radiation and the willingness to buy III

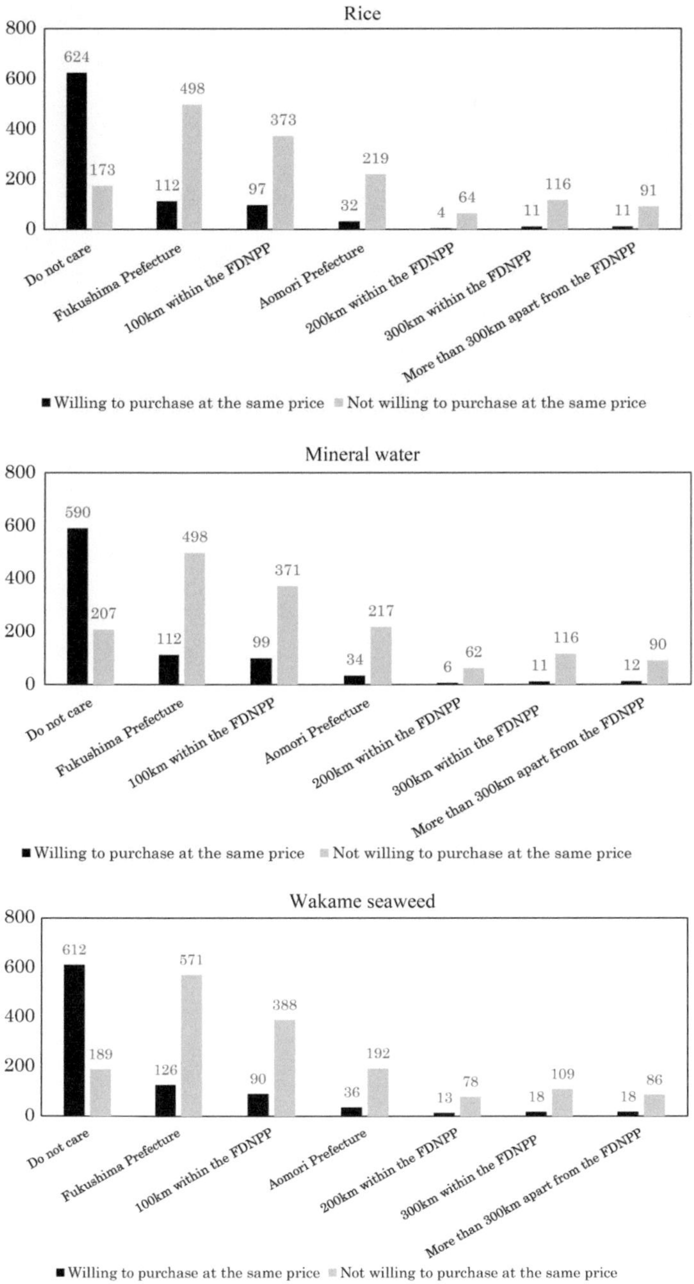

Fig. 5.56 Perceptions of product origin and the willingness to buy III

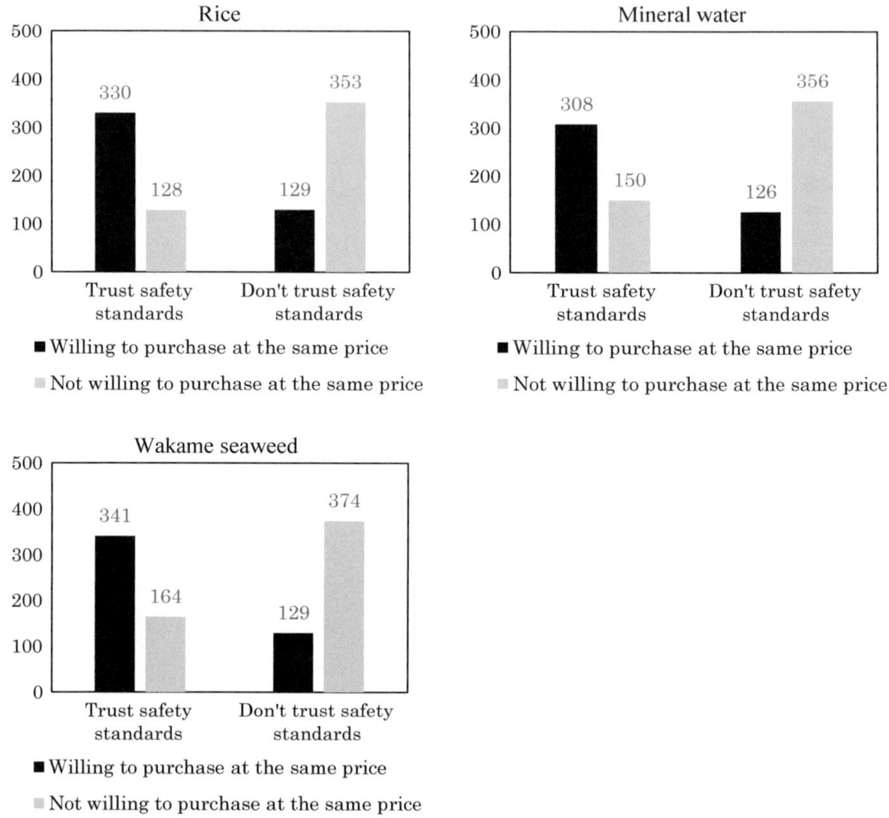

Fig. 5.57 Trust toward food safety standards for radiation and the willingness to buy III

You can see from the figure that there is a large disparity between the respondents who trust and do not trust the food safety standards. It is evident that respondents with trust in the food safety standards have a positive reaction toward food and beverage from these regions. On the contrary, those without trust in the safety standards have a negative reaction toward the products.

Thus the respondents' levels of trust in the food safety standards have a positive effect on their willingness to buy rice, mineral water, and wakame seaweed from regions near the FDNPP.

5.4.5 Respondents' Perceptions of Willingness to Accept Food from Regions Near the FDNPP and Their Willingness to Buy

In this subsection, I would like to discuss how the respondents' perceptions of willingness to accept rice, mineral water, and wakame seaweed from regions near the FDNPP are different among the products and how having safety labels on the products will affect the respondents' willingness to buy.

Respondents' willingness to accept perceptions for rice, mineral water, and wakame seaweed

Table 5.10 describes the percentages of respondents that are willing to and not willing to accept rice, mineral water, and wakame seaweed from regions near the FDNPP at different levels of price discounts. From the table, you can see that the percentage of respondents that are willing to accept food and beverage products from regions near the FDNPP at no price discount is lower compared to the foods we investigated in the previous subsections. This result is expected because the products of our interest in this subsection are those that a majority of the respondents are not willing to buy products from regions near the FDNPP without any discounts.

It is also apparent in the table that percentage of respondents who avoid buying products of regions near the FDNPP even with a 60% price discount is higher compared with the foods of previous subsections. In particular, over one-third of the

Table 5.10 Respondents' willingness to buy rice, mineral water, and wakame seaweed

	Number of samples	Percentages %
Rice		
Willing to buy at the same price	834	48.8
Willing to buy when the discount rate is between 10 and 60%	313	18.3
Not willing to buy even with a 60% discount rate	563	32.9
Mineral water		
Willing to buy at the same price	798	46.7
Willing to buy when the discount rate is between 10 and 60%	264	15.4
Not willing to buy even with a 60% discount rate	648	37.9
Wakame seaweed		
Willing to buy at the same price	817	46.2
Willing to buy when the discount rate is between 10 and 60%	368	20.8
Not willing to buy even with a 60% discount rate	583	33.0

respondents for mineral water are not willing to accept such product even if there is a 60% price discount. This suggests that the degree of concern regarding the risk of radioactive contamination is very high for this product. The reason for this is perhaps because a high level of radioactive cesium is detected in not only the drinking water of Fukushima, Miyagi, and Ibaraki prefectures, but also in Tokyo and Kanagawa prefectures after the Fukushima disaster. Consequently, numerous individuals realize the radioactive contamination of water is a severe problem. Therefore, it could be that many consumers recognize the risk of the radioactive contamination of water as not just another false information spread by media but is an actual problem.

The effects of having safety labels on food products

Figure 5.58 reveals the percentages of respondents that are willing to and not willing to buy rice, mineral water, and wakame seaweed from regions near the FDNPP if there are safety labels on the products showing that they have no risk of

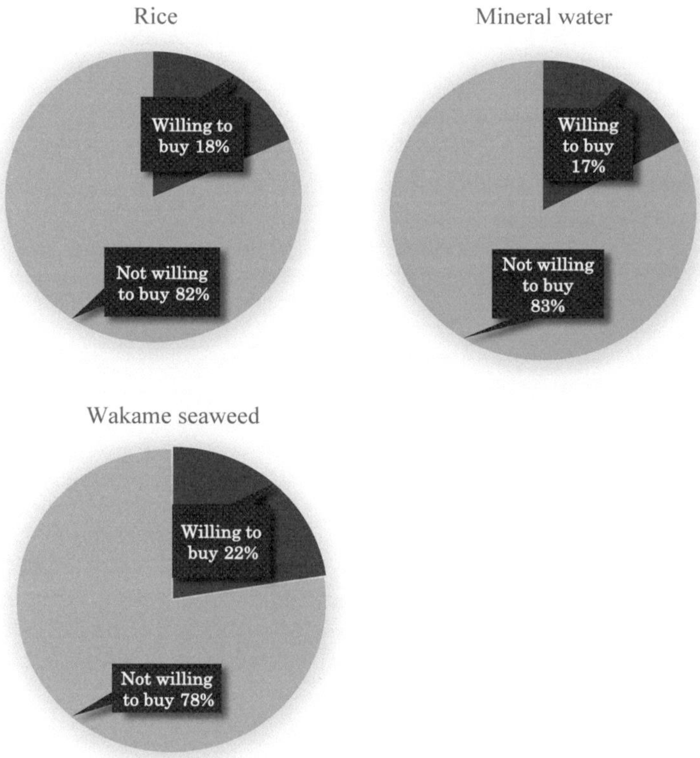

Fig. 5.58 Food label that shows the product is safe from radiation and the willingness to buy III

radioactive contamination. As also explained in the previous subsections, the percentages in the figure are those of the respondents who are not willing to buy products from regions near the FDNPP even at a 60% price discount.

For rice and mineral water, it can be seen from the figure that more than 80% of the respondents will not change their decisions even after safety labels are placed on the products. This percentage is higher among the ten products of our interest, which indicates that it is difficult to restore the sales for these products by transmitting information on product safety through labeling.

On the other hand, the respondents who will not change their decisions about purchasing wakame seaweed are less than 80%, which suggests that at least more than one-fifth of the respondents are likely to buy wakame seaweed from regions near the FDNPP if there are safety labels on the products. Hence, placing safety labels on products is more effective for wakame seaweed compared with rice and mineral water.

5.4.6 Respondents' Social Attributes and Their Willingness to Buy

Finally, as the last part of this subsection, I will discuss how respondents' social attributes have effects on their willingness to buy rice, mineral water, and wakame seaweed from regions near the FDNPP. We will go over Figs. 5.59 to 5.65 for this purpose.

Respondents' residential distance from the FDNPP and the willingness to buy

First, Fig. 5.59 explains how differences in the respondents' residential distance from the FDNPP have an influence on their willingness to buy.

For rice and mineral water, you can see that respondents who reside within the 300 km distance from the FDNPP are more likely to respond positively about buying products from regions near the FDNPP. However, when the distance exceeds 300 km, more respondents react negatively toward these products.

For wakame seaweed, respondents start to respond negatively when there residential distance from the FDNPP exceeds 100 km, demonstrating that more of the respondents have tendencies to avoid buying wakame from regions near the FDNPP.

Among the respondents' living more than 400 km apart from the FDNPP, more than a half of the respondents have a negative reaction about buying the products in all three products.

Therefore, residential distance has a negative impact on the respondents' willingness to buy products from regions near the FDNPP. The reason for this is that rice, mineral water, and wakame seaweed are all produced in various areas besides the regions near the FDNPP and can also be imported so that the consumers can easily find substitutes to avoid products from regions near the FDNPP.

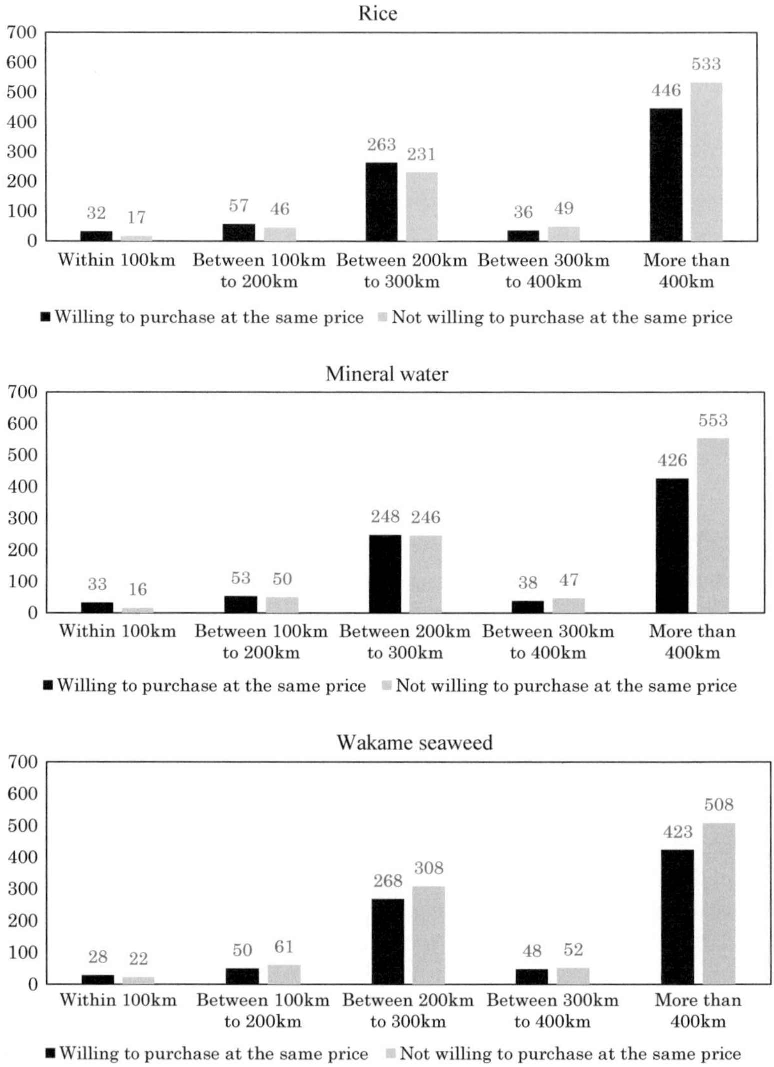

Fig. 5.59 Location of residence and the willingness to buy III

Respondents' gender and their willingness to buy

Next, in Fig. 5.60, I would like to seek how the respondents' willingness to buy rice, mineral water, and wakame seaweed from regions near the FDNPP are different between their genders.

From the figure, it is apparent that the reactions toward rice and mineral water from regions near the FDNPP are different between the genders. Male respondents are more likely to agree about buying rice and mineral water from these regions, but

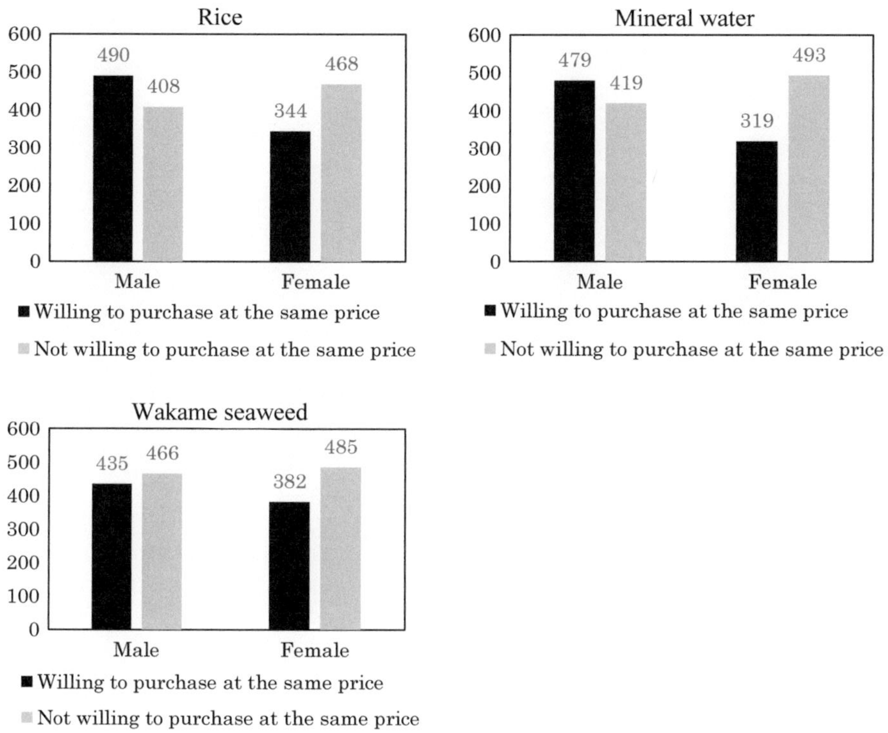

Fig. 5.60 Gender and the willingness to buy III

female respondents have an adverse reaction toward these products. This gap in the reaction between the genders are similar to the results in the previous subsections, but for wakame seaweed, both male and female respondents show a negative response about buying wakame seaweed from regions near the FDNPP. Therefore, it is believable that the consumers are highly concerned about the risk of radioactive contamination of wakame seaweed.

Respondents' age and their willingness to buy

Figure 5.61 depicts the relationship between the respondents' age and their willingness to buy three foods of our interest.

Here too, it is evident that younger respondents are more likely to have a negative reaction toward food and beverage produced in regions near the FDNPP.

In particular, respondents in the 30s and 40s have a very high avoidance rate compared with other age groups. As discussed previously, the reason for this is respondents in these age groups are the ones who are raising little children and are afraid of the risk of their children being exposed to radiation.

It is also seen in the figure that respondents in the 60s have a positive willingness to buy reaction suggesting that the respondents' willingness to buy becomes higher as their age increases.

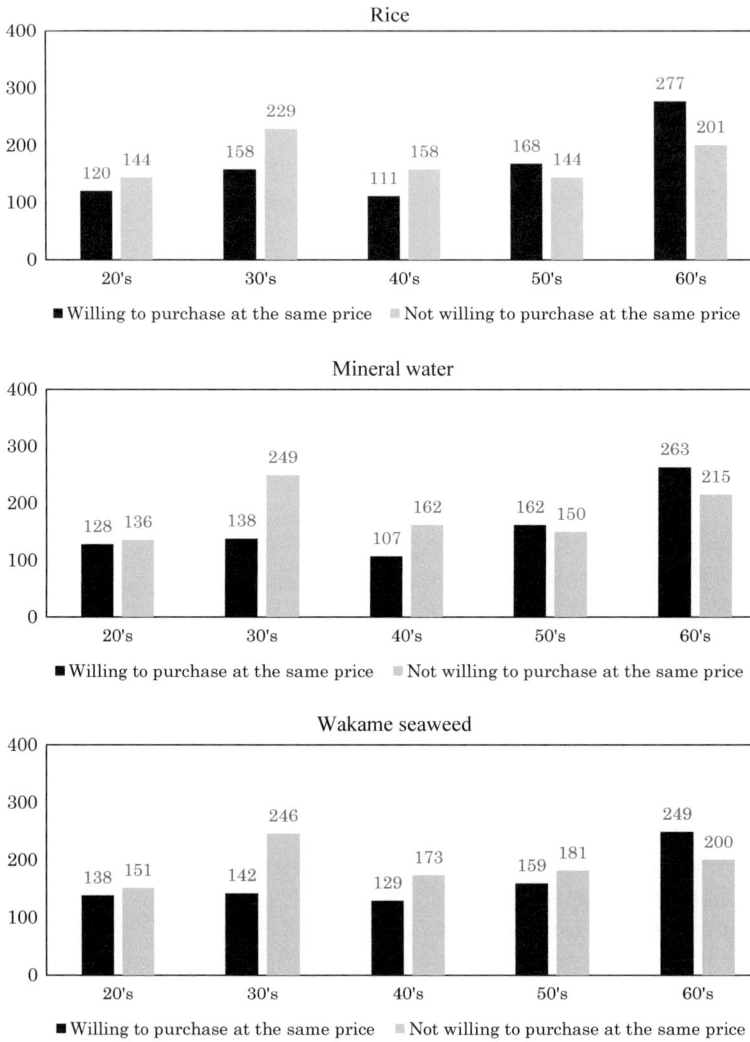

Fig. 5.61 Age and the willingness to buy III

Relationships between the presence or absence of children and the willingness to buy

Here I would like to compare the willingness to buy rice, mineral water, and wakame seaweed from regions near the FDNPP between the respondents with children and those without children through Figs. 5.62 and 5.63.

According to Fig. 5.62, in all products, respondents with infants and elementary school age children have the tendency to avoid buying products from regions near the FDNPP.

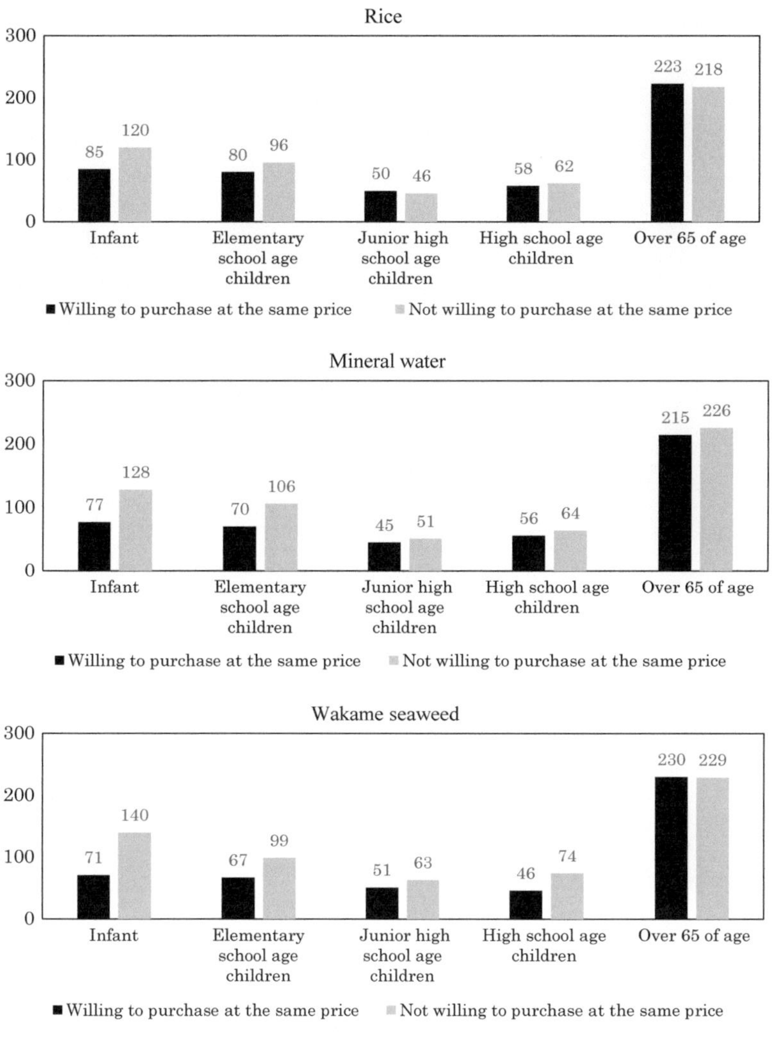

Fig. 5.62 Family structure and the willingness to buy III

It can also be seen from the figure that respondents living with a family member of age over 65 are showing a relatively positive reaction toward rice and wakame seaweed from regions near the FDNPP. However, for mineral water, even respondents living with a family member of age over 65 are responding negatively about buying the product. This suggests that the consumers are very worried about the risk of purchasing radioactively contaminated mineral water.

Furthermore, Fig. 5.63 shows that in all three products respondents with more than two children are more likely to avoid products from regions near the FDNPP.

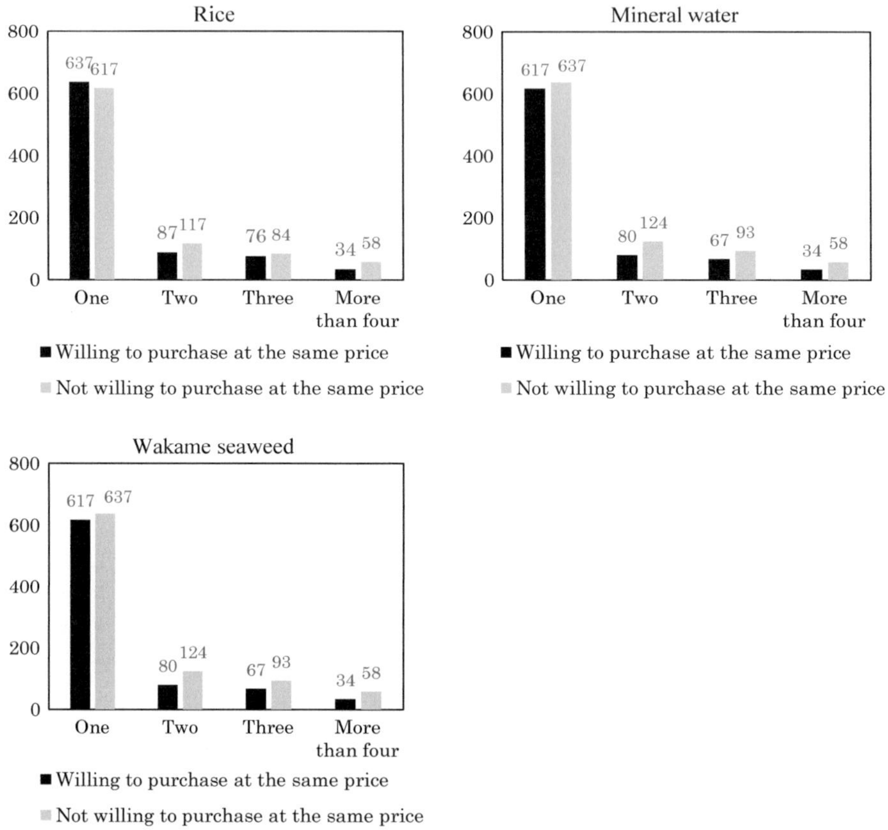

Fig. 5.63 Number of children under age 15 and the willingness to buy III

For mineral water and wakame seaweed, even respondents with one child have a negative reaction toward these products.

From the above investigation, it is presumable that the consumers are highly concerned about the risk of their children getting exposed to radiation from purchasing rice, mineral water, and wakame seaweed.

Educational level and the willingness to buy

Figure 5.64 presents how respondents' educational levels have impacts on their willingness to buy rice, mineral water, and wakame seaweed from regions near the FDNPP.

You can see from the figure that more of the respondents with an educational level of Junior college or Bachelor's degree are not willing to buy the products compared with those with a high school degree. This suggests that consumers with a higher educational level are evaluating highly of the risk of products being

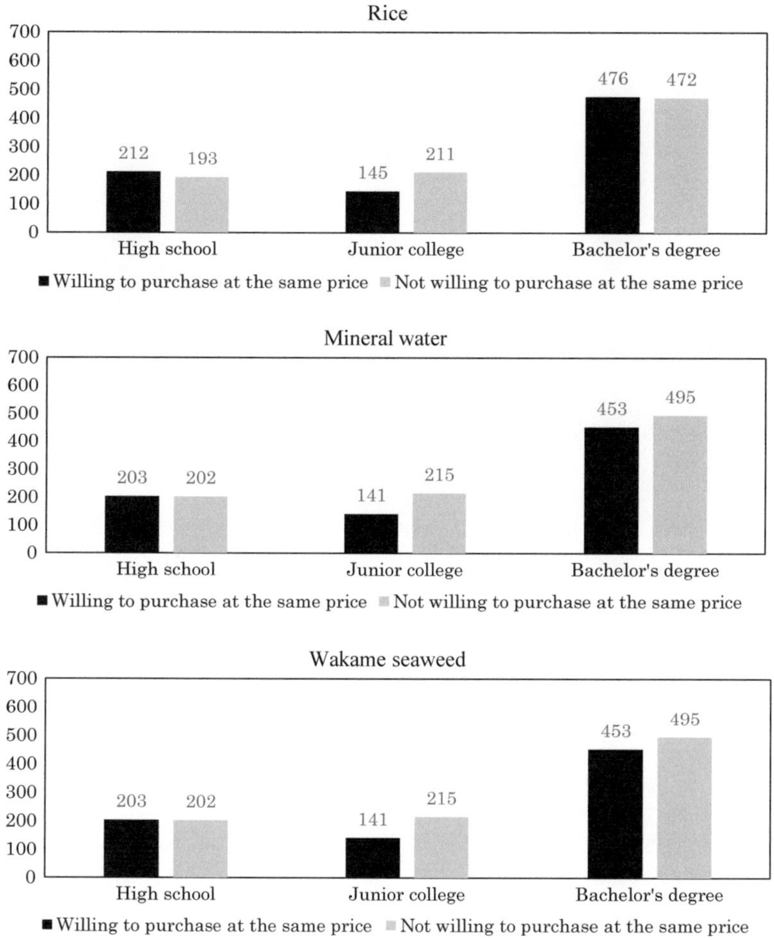

Fig. 5.64 Highest educational level and the willingness to buy III

contaminated with radiation and that they are more likely to avoid rice, mineral water, and wakame seaweed from regions near the FDNPP. As explained in the previous subsection, it is conceivable that the reason for this is because consumers with higher educational levels are more knowledgeable about the relations between the radiation dose and the risk of becoming cancer.

Respondents' income level and the willingness to buy

As the last part of this subsection, I would like to go over in Fig. 5.65 the relationship between the respondents' income level and their willingness to buy rice, mineral water, and wakame seaweed from regions near the FDNPP.

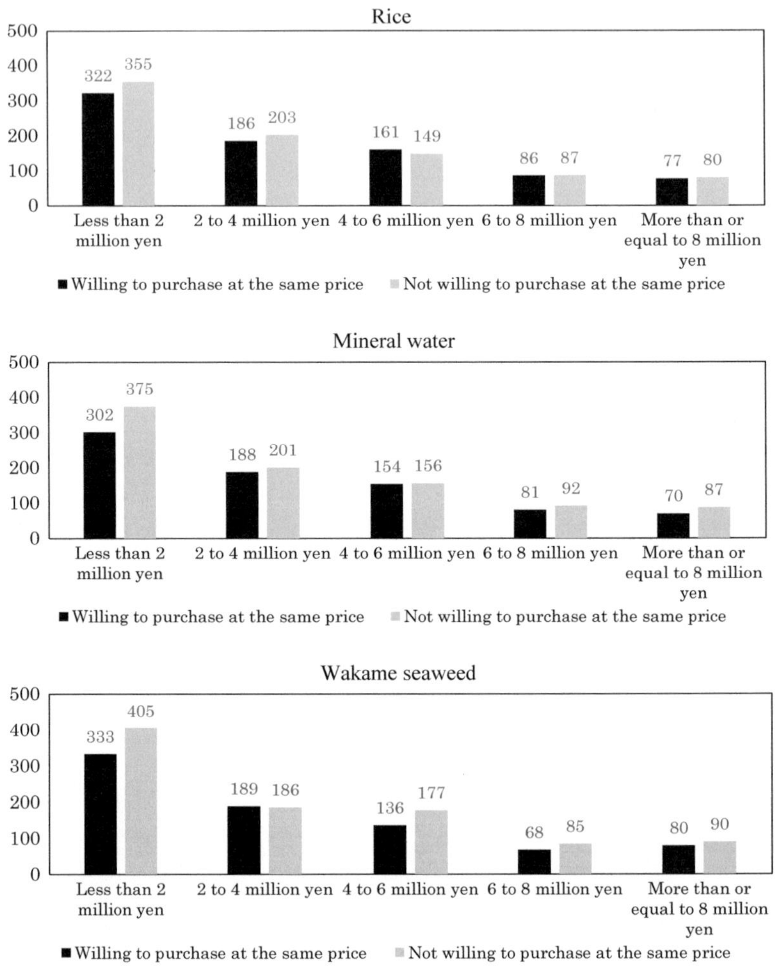

Fig. 5.65 Income level and the willingness to buy III

From the figure, it is evident that in almost all the income groups more of the respondents are not willing to buy products from regions near the FDNPP. Furthermore, it is apparent that respondents with an annual income of fewer than 2 million yen have a high avoidance rate toward buying these products. It is likely that the reason for this is because more respondents that belong to this income group are housewives, females, and have high cooking frequencies, which leads to a low willingness to buy.

Hence, from the figure, regardless of the income levels, a majority of the respondents are reacting negatively about buying the three products of our interest in this subsection. This implies that among the ten products investigated in this book, respondents have the highest concern about the risk of radioactive contamination for rice, mineral water, and wakame seaweed produced in regions near the FDNPP.

5.5 Factors Affecting the Consumers' Willingness to Buy Products from Regions Near the FDNPP

As a summary of this chapter, I would like to explain the factors that are likely affecting the consumers' willingness to buy products from regions near the FDNPP.

In this chapter, we mainly looked into how the following six factors have an influence on the consumer reaction toward products from regions near the FDNPP. The first factor is the consumers' eating habits, the second is their perceptions of food safety, and the third is their interests in social problems. Furthermore, the fourth factor is the consumers' attitude toward nuclear contamination, and the fifth is their willingness to accept perceptions, and the final one is their social attributes.

In this section, we will reconfirm the factors that are likely affecting the consumers' willingness to buy through Tables 5.11 to 5.13. These tables show the summary of our findings from the analyses we performed in this chapter.

5.5.1 Effects of Respondents' Eating Habits and Food Safety Perceptions

Table 5.11 summarizes the results of the effects of the respondents' eating habits and food safety perception on their reaction toward products from regions near the FDNPP. The minus and plus signs in the table denotes that the respondents' have a negative and a positive willingness to buy products of these regions. "No effects" in the table means that it was not possible to confirm from our investigation whether the effects had a positive or a negative impact.

Eating habits

First, it can be seen from the figure that consumers with a high shopping and cooking frequency, and put emphasis on safety issue when buying food products have a negative influence on their willingness to buy. The reason for consumers with a high shopping and cooking frequency to avoid products from regions near the FDNPP is perhaps because such consumers have more opportunities to buy and use food products and are more likely to highly evaluate the risk of radioactive contamination of food products. Furthermore, it is presumable that consumers who put emphasis on safety issue have a negative willingness to buy because such consumers are highly concerned about the risk of buying radioactively contaminated food from regions near the FDNPP.

Table 5.11 Effects of respondents' eating habits and food safety perceptions

	Eating habits						Perception of food safety issue	
	Shopping frequency	Shopping experiences	Cooking frequency	Eating style	Emphasis on safety	Emphasis on price	Cares about safety issue	Trusts on food labels
Products that a majority of the respondents are willing to buy	–	No effects	–	–	–	+	–	+
(Cucumber, apple, beef, and pork)	Beef and pork	All products	Pork	Beef and pork (who often go out to eat)	Apple, beef, and pork	All products	Very apparent in beef and pork	All products
Products that about a half of the respondents are willing to buy	–	+	–	–	–	+	–	+
(Shiitake mushrooms, chicken eggs, and tuna)	Tuna	All products	All products	Shiitake mushrooms and chicken eggs (who usually eat at home)	All products	All products	All products	All products
Products that a majority of the respondents are not willing to buy	–	No effects	–	No effects	–	–	–	+
(Rice, mineral water, and wakame seaweed)	All products	All products	All products	All products	All products	All products	All products	Rice

From the column for the "shopping experiences" in Table 5.11, it is evident that consumers with experience in buying shiitake mushrooms, chicken eggs, and tuna fish are more likely to have a positive willingness to buy. By contrast, consumers with no experience in buying such foods are not willing to buy these foods from regions near the FDNPP.

From the fourth column of the table, although some negative effects are apparent among the consumers who often eat their meals at home, the outcomes have been different among the ten products. Thus, it is not possible to conclude that consumers' difference in the eating style (whether they eat at home or outside) have impacts on their reaction toward the products from these regions.

Finally, it is discernible from the sixth column of the table that reaction among the consumers who put importance on price when buying food at stores have an inconsistent result among the ten products. As seen from the figure, for products that a majority and about a half of the respondents are willing to buy, consumers that put emphasis on price are more willing to purchase products from regions near the FDNPP. On the other hand, among the respondents for products that a majority of the respondents are not willing to buy, respondents that are concerned about the price have a negative reaction about buying products from such regions. Hence, it is probable that for rice, mineral water, and wakame seaweed, discounting the price of these products will not change the consumer reaction toward these products.

Food safety perceptions

Next, I would like to discuss how the respondents' perceptions of food safety issue are related to their willingness to buy. From the column "Cares about safety issue" of Table 5.11, respondents who take action to avoid unsafe food have a negative response about buying these products. It is likely that consumers who are highly conscious about food safety issues are more concerned about the risk of products from regions near the FDNPP to contain radioactive materials.

By contrast, you can see the last column of Table 5.11 that respondents who have a trust in the information they receive from food labels are more willing to buy products from regions near the FDNPP. This implies that improving the reliability of information placed on food labels can help mitigate the consumers' fear toward the food produced in these regions.

5.5.2 *Effects of Respondents' Interests in Social Problems and Risk Perceptions Toward Radioactive Contamination*

Table 5.12 shows how the respondents' interests in social problems and risk perceptions toward radioactive contamination have an influence on their willingness to buy products from regions near the FDNPP. The minus and plus signs in the table

Table 5.12 Effects of respondents' interests in social problems and risk perceptions of radioactive contamination

	Interests in social problems		Perceptions of radioactive contamination					
	Enthusiastic about environmental activities	Enthusiastic about helping in the regions damaged by the Fukushima disaster	Risk of food being contaminated with radio active material	Knowledge of radiation			Concerned about the product origin	Have a trust in the safety standards for radio active cesium
				Radiation do se and the risk of developing cancer	Knowledge other than developing cancer	No Knowledge		
Products that a majority of the respondents are willing to buy	+	+	–	–	+	–	–	+
(Cucumber, apple, beef, and pork)	All products	All products	All products	Beef and pork	All products	All products	All products	All products
Products that about a half of the respondents are willing to buy	+	+	–	–	+	–	–	+
(Shiitake mushrooms, chicken eggs, and tuna)	All products	All products	All products	All products	All products	All products	All products	All products

(continued)

Table 5.12 (continued)

	Interests in social problems		Perceptions of radioactive contamination					
	Enthusiastic about environmental activities	Enthusiastic about helping in the regions damaged by the Fukushima disaster	Risk of food being contaminated with radio active material	Knowledge of radiation			Concerned about the product origin	Have a trust in the safety standards for radio active cesium
				Radiation do se and the risk of developing cancer	Knowledge other than developing cancer	No Knowledge		
Products that a majority of the respondents are not willing to buy	+	+	–	–	Different results among the type of knowledge	–	–	+
(Rice, mineral water, and wakame seaweed)	All products	All products	All products	All products	All products All products	All products	All products	All products

again represents that the factors of our interest have negative and positive effects on the consumers' willingness to buy.

First, it is clear from the "Interests in social problem" column that in all products consumers with a high interest in social problems have a positive willingness to buy products from regions near the FDNPP. As explained before, having this positive effect on the consumers' willingness to buy is likely because respondents that are interested in social problems such as preserving the environment or helping the disaster areas to recover from their damage are more enthusiastic about helping other people. Therefore, consumers with a high interest in social problems tend to be altruistic consumers. It is likely that appealing to such consumers about the effectiveness of purchasing products from regions near the FDNPP on their economy will help them recover from their damage.

Second, I would like to have an overview of the effects of respondents' perceptions regarding the risk of radioactive contamination upon their willingness to buy reactions. According to Table 5.12, you can see that respondents that consider highly of the risk of food being contaminated with radioactive material and have knowledge about the relations between radiation dose and developing cancer have a negative reaction toward products from regions near the FDNPP. Furthermore, those that know nothing about radiation and are concerned about the product origin also react negatively about buying products from these regions. On the other hand, knowledge of radiation other than the risk of developing cancer such as the types of radiation rays and radioactive isotopes decay until becoming stable isotopes have a positive impact on the consumers' willingness to buy. Furthermore, respondents' trust toward the food safety standards for radiation also has a positive effect.

This result tells us that factors that lead to evaluating highly of the risk of radioactive contamination have an adverse effect on the willingness to buy. However, factors that help lower the consumers' risk perceptions of radioactive contamination such as knowledge of radiation and credibility of the food safety standards lead to a positive consumer reaction.

5.5.3 Effects of Respondents' Perceptions of Willingness to Accept and Social Attributes

Table 5.13 encapsulates the results on how the respondents' perceptions of willingness to accept and social attributes are related to their reactions toward products of regions near the FDNPP.

Willingness to accept products from regions near the FDNPP

We found from our survey that quite a large number of respondents are not willing to accept products from regions near the FDNPP even when there is a 60% discount on the price. Thus, we asked these respondents whether they would accept buying these products if there is a label on the product showing that the products are free from radioactive contamination. You can see from Table 5.13 that the effects of

Table 5.13 Effects of respondents' willingness to accept food produced in regions near the FDNPP and social attributes

	Willingness to accept food produced in regions near the FDNPP	Social attributes						
	Reaction toward food labels	Distance of the residence	Gender	Age	Family structure	Number of children	Education	Income
Products that a majority of the respondents are willing to buy	Have effects on about 20% of the people who were not willing to accept the products without the label		Male: + Female: −	Age under 40: −	Household with children under 12: −		No effects	Less than 2 million yen and above 8 million yen: −
(Cucumber, apple, beef, and pork)	All products	Beef and pork	All products	All products	All products	All products	All products	Pork
Products that about a half of the respondents are willing to buy	Have effects on about 21–25% of the people who were not willing to accept the products without the label		Male: + Female: −	Age under 30: −	Household with children under 12: −			Between 4 million and 8 million yen: −
(Shiitake mushrooms, chicken eggs, and tuna)	All products	All products	All products	All products	All products	All products	All products	All products
Products that a majority of the respondents are not willing to buy	Have effects on about 20% of the people who were not willing to accept the products without the label		Male: + Female: −	Age under 50: −	Household with children under 12: −			No effects
(Rice, mineral water, and wakame seaweed)	Wakame seaweed	All products	Rice and mineral water	All products	All products	All products	All products	All products

placing safety labels only have a limited effect on the consumers' willingness to accept products from regions near the FDNPP. As seen in the table, only about 20% of the respondents with high avoidance rate change their response about buying these products.

In particular, among the products that a majority of the respondents are not willing to buy, only wakame seaweed had more than 20% of such respondents to change their willingness to accept responses. Furthermore, for rice and mineral water, the respondents with high avoidance rate continued to react negatively about buying these products even after radiation safety labels are placed on them.

Hence, when the respondents are very anxious about the risk of radioactive contamination in food and beverage products, putting safety labels on products only has a limited effect on the consumers' willingness to accept products from regions near the FDNPP.

Social attributes

Next, I would like to describe the results of how respondents' social attributes had impacts on their willingness to buy products from regions near the FDNPP. You can see from Table 5.13 that the distance of the respondents' residence from the FDNPP, age groups with little children, who have children under 12 years old, number of children, and the respondents' educational level have a negative influence on the willingness to buy.

First, the reason that the respondent's residential distance have a negative impact is likely because respondents' that live farther from the FDNPP have a fewer chance to see these products in their local grocery stores and do not have much information regarding the safety of the products.

Second, age, family structure, and number of children have a negative effect because these factors are related to whether the respondents have contacts with children such as they were in age groups raising children, have children under age 12, and live with more than two children. As it is known that children are highly susceptible to radiation compared with adults when exposed to radioactive material, it is conceivable that consumers having more contacts with children wanted to avoid their children from having food with a risk of radioactive contamination.

The results of the effects of the respondents' gender have an inconsistent outcome between the male and female respondents. While male respondents are more likely to show a positive reaction toward products of regions near the FDNPP, female respondents have a negative attitude about buying products from these regions. It is conceivable that women spend more time with their children than males and that they have the tendency to avoid buying food from these regions because they want to prevent their children from the risk of radiation exposure.

Educational levels have a negative impact on the consumers' willingness to buy although no effect is determined for the products that a majority of respondents are willing to buy. The reason for the negative impact is that individuals with higher educational levels are more likely to know about the relations between the radiation dose and the risk of becoming cancer.

Finally, we do not find much evidence on whether the respondents' income levels have an influence on their willingness to buy. Hence, I would like to conclude that income does not have a specific effect on the consumers' reaction toward the products of regions near the FDNPP.

References

Aruga, K. (2016). Consumer responses to food produced near the Fukushima nuclear plant. *Environmental Economics and Policy Studies*. doi:10.1007/s10018-016-0169-y

Cucinotta, F. S., & Durante, M. (2009). Risk of radiation carcinogenesis. In J. C. McPhee & J. B. Charles (Eds.), *Human health and performance risks of space exploration missions* (pp. 119–170). Houston: NASA Johnson Space Center.

Kolstad, C. D. (2000). *Environmental economics*. New York: Oxford University Press.

Ministry of Internal Affairs and Communications (MIC). (2013). 2013 Population Census (in Japanese). http://www.stat.go.jp/data/jinsui/2013np/. Accessed 24 Feb 2017.

Chapter 6
Is There Reputation Damage?

In this chapter, I would like to investigate whether the six factors we investigated in the previous chapter, consumers' eating habits, perceptions of food safety, interests in social problems, attitudes toward nuclear contamination, perceptions of willingness to accept, and social attributes are causing reputation damage to regions near the FDNPP. Then, I will find out if food and beverage products of these regions are being affected by bad reputation.

To provide the answer to the question, "Is there reputation damage," we find that there are cases that the reputation is a major factor causing damages to the products in regions near the FDNPP. Yet, there are also circumstances where factors not directly related to reputation are contributing to consumers' negative attitudes toward products of these regions.

Hereafter, I would like to explain this finding based on the two cases. The first case is that one of the six factors we investigated in this book is causing reputation damage. The second case is that other factors that are not directly related to false rumors are affecting the consumers' reaction toward products of these regions. We will discuss these cases using Table 6.1.

6.1 Eating Habits and Reputation Damage

First, I would like to describe the factors within the consumers' eating habits that are likely related to reputation damage and those that are not directly related to reputation damage. In the following subsections, I would like to explain the details of these factors. The summary of the discussion is provided in Table 6.1.

K. Aruga, *Consumer Reaction, Food Production and the Fukushima Disaster*,
DOI 10.1007/978-3-319-59849-9_6

Table 6.1 Factors affecting negatively on the willingness to buy

	Factors related to reputation damage	Factors not directly related to reputation damage
Factors related to eating habits	Consumers that consider the price is important do not have a positive willingness to buy even when there are price discounts	Consumers with a high cooking frequency have a lower willingness to buy
Factors related to perceptions of radioactive contamination	Consumers without radiation knowledge have a lower willingness to buy	Consumers with knowledge about the relations between the radiation dose and developing cancer have a lower willingness to buy
Factors related to social attributes	Consumers living farther from the FDNPP have a lower willingness to buy	Consumers having little children have a lower willingness to buy

6.1.1 The Factor Causing Reputation Damage

In Q6 of Table 5.1, we asked the consumers about what they are concerned about the most when they go grocery shopping, and many of them answered that they put emphasis on the price of products. Consumers who selected price as what they put emphasis on when buying products are the ones who are sensitive to the price change. Hence, such consumers are more likely to change their purchase decisions when prices change. However, we found that even among such price-sensitive consumers, most of them did not react to the price discounts provided for the products of regions near the FDNPP.

Table 6.2 shows the percentage of consumers that consider the price is important when buying products in grocery stores, but is reluctant about buying products even with a 60% price discount on the products from regions near the FDNPP. 60% price discount was the highest price discount that we had in our survey questionnaire but as seen in the table more than one-fourth of the consumers who put importance on price when buying products have contradictive responses toward products from these regions. When asking the consumers' willingness to buy the products, we had supplementary information that the products from these regions meet the national food safety standards for radiation (see Q16 of Table 5.3). However, many consumers continued to reveal a negative attitude as we increased the levels of price discounts for the products.

This adverse reaction toward products from regions near the FDNPP among the consumers that are sensitive to a price change is not a rational response and it is presumable that these consumers have been somewhat affected by a false rumor or reputation. In particular, for rice, mineral water, and wakame seaweed, nearly or more than one-third of the price-sensitive consumers are avoiding products from these regions and it is believable that some of the behaviors of these consumers are highly related to the effect of a bad reputation, leading to reputation damage.

Table 6.2 Percentages of consumers that consider price is important but are reluctant about buying products even with a 60% price discount

	Products	Percentage (%)
Products that a majority of the respondents are willing to buy	Cucumbers	26.4
	Apples	25.3
	Beef	26.0
	Pork	25.4
Products that about a half of the respondents are willing to buy	Shiitake mushrooms	29.6
	Chicken eggs	27.8
	Tuna fish	28.4
Products that a majority of the respondents are not willing to buy	Rice	32.5
	Mineral water	37.8
	Wakame seaweed	33.7

6.1.2 The Factor not Directly Causing Reputation Damage

Here, I would like to explain the factor of consumers' eating habits that is not directly causing reputation damage but are affecting negatively on the consumers' reaction toward the products from regions near the FDNPP.

As seen in Table 5.11 of Chap. 5, although shopping and cooking frequencies have a negative effect on the consumers' willingness to buy, it is conceivable that the reason that consumers avoid buying food from these regions is not only because they are affected by false rumors regarding those products, but also because they are worried about the cumulative effect of radiation.

The current Japanese food safety standards for radioactive cesium is relatively a severe standard because when the products meet this standard, it will mean that the effect of the radioactivity on the individual's body is limited within 1 mSv even if the individuals consumed the product every day. Hence, the Japanese food standard for radioactive cesium does take account of the cumulative effect of radiation. However, there is a gap within the individuals regarding the amount of food they eat, and there may be individuals having unbalanced diets and eating certain products many times a day. If so, there could be some cases that the national food safety standards do not apply. Therefore, the avoiding behavior of consumers with such abnormal eating habits is not only attributed to the effects of reputation. It is more likely that such behavior is based on the individual's rational decision to avoid and mitigate the cumulative effect of radiation.

6.2 Knowledge of Radiation and Reputation Damage

Second, I would like to see among the factors related to consumers' knowledge of radiation that is likely causing reputation damage and those that is not directly causing such damage to regions near the FDNPP. The summary of our discussion of this subsection is provided in Table 6.1.

6.2.1 The Factor Causing Reputation Damage

As we found in the previous chapter, consumers that do not possess any knowledge of radiation have a negative reaction toward products of regions near the FDNPP. It is believable that consumers without knowledge of radiation are susceptible to reputation damage, because these consumers cannot judge correctly or are indifferent about the truthfulness of reputation regarding the risk of radioactive contamination of products from regions near the FDNPP. Hence, it is likely that the negative reaction of such consumers is somewhat influenced by false rumors and thus the absence of radiation knowledge can be a cause of reputation damage.

Table 6.3 illustrates the percentage of respondents among those with and without knowledge of radiation that is willing to buy products from regions near the FDNPP at the same price as those from regions farther from the FDNPP. Respondents with radiation knowledge are those who knew any of the knowledge of radiation provided in Q12 of Table 5.2. According to the table, respondents with radiation knowledge are more likely to accept buying products from regions near the FDNPP compared with those without radiation knowledge. This indicates that respondents without radiation knowledge are reacting more negatively toward products from these regions. It is presumable that the reason for this gap between the reaction of respondents with and without radiation knowledge is because respondents without radiation knowledge are more inclined to be affected by bad rumors and consider products from regions near the FDNPP as products with risk of radioactive contamination. Hence, it is conceivable that the behaviors of some of the consumers without radiation knowledge are contributing to reputation damage.

Table 6.3 Radiation knowledge and the willingness to buy

	Percentage of respondents willing to buy at the same price (%)
With radiation knowledge	52.9
Without radiation knowledge	44.9

6.2.2 The Factor not Directly Causing Reputation Damage

Next, I would like to explain the factors related to consumers' knowledge of radiation that is not necessarily having direct effects on the reputation of products from regions near the FDNPP.

Among the radiation knowledge that has a negative impact on the consumers' reaction toward products from these regions, it is believable that the negative reaction of consumers with knowledge of radiation dose and the risk of becoming cancer is not related to the effects of reputation. This is because consumers who know that "when an additional amount of radiation received exceeds 100 mSv the probability of developing cancer in a lifetime increases" are avoiding products of regions near the FDNPP based on their own knowledge about the risk of radioactive contamination. Therefore, their avoiding behavior is not only the effect of false rumors.

All in all, it is likely that the behavior of consumers with knowledge of radiation dose and the risk of becoming cancer is not directly causing reputation damage.

6.3 Social Attributes and Reputation Damage

Third, let us seek the factor among the social attributes having a negative influence on the consumers' willingness to buy that is and is not leading to reputation damage. Likewise with the previous subsections, the summary of the argument is shown in Table 6.1.

6.3.1 The Factor Causing Reputation Damage

As explained in the previous chapter, consumers living farther from the FDNPP have a negative reaction toward products from regions near the FDNPP compared with those living near the FDNPP. This finding is likely contributing to reputation damage.

We saw in the previous chapter that consumers have a positive reaction about buying products from regions near the FDNPP when products are accessible in all parts of Japan such as cucumbers and apples. However, for products that are not distributed countrywide and thus have low accessibility, consumers have a negative reaction when they are from regions near the FDNPP. In particular, it was apparent that consumers with a longer residential distance from the FDNPP were more likely to avoid products from these regions compared with those with a shorter distance from the FDNPP. The reason for this disparity in the willingness to buy between the consumers with a longer and a shorter residential distance from the FDNPP is

Table 6.4 Residential distance from the FDNPP and the percentage of respondents not willing to buy products with a 60% price discount

Distance from the FDNPP	Percentage (%)
Less than 100 km	20.9
100–200 km	31.0
200–300 km	29.5
300–400 km	29.8
More than 400 km	29.6

perhaps because consumers who live farther from the FDNPP have little chance to see products from regions near the FDNPP in their local grocery stores.

Thus, it is probable that consumers living apart from the FDNPP are avoiding products from regions near the FDNPP, because they lack information about the product and are somewhat influenced by false rumors. What makes this situation worse is that agricultural products have many substitutes so that the consumers do not need to confirm the truthfulness of the information regarding products of these regions. This is why the conditions of reputation damage of regions near the FDNPP can get worse.

Table 6.4 presents the percentage of respondents not willing to buy products from regions near the FDNPP within the respondents' group with different residential distance from the FDNPP. The table is created by including all the survey respondents. As shown in the table, about 20% of the respondents that live less than 100 km from the FDNPPP have a negative reaction toward products of regions near the FDNPP, while this percentage of respondents avoiding products from these regions become nearly 30% when their residential distance is more than 100 km from the FDNPP. Hence, here too it is evident that respondents who live farther from the FDNPP are more likely to avoid buying products from these regions compared with those that live near the FDNPP. Again, it is credible that this is because consumers that live farther from the FDNPP lack information about the products of regions near the FDNPP and are more susceptible to the effects of false reputation.

6.3.2 The Factor not Directly Causing Reputation Damage

On the other hand, it is believable that factors among the social attributes such as female, have little children in their households to have a negative impact on the consumers' willingness to buy are not likely caused by the effects of false rumors. This is because the reason for female consumers and consumers with little children to avoid buying products from regions near the FDNPP is somewhat related to consumers' radiation knowledge that children are more at risk of developing tumors when exposed to radiation. Reacting negatively about buying products from these regions to prevent their children from the risk of having food containing radioactive cesium is a rational decision in a sense because the national food safety standards are not omnipotent when the cumulative effect of radiation is considered.

In this sense, the consumers' negative reaction is partly attributed to these above mentioned social attributes, and hence, such reaction is not only caused by the effect of a false rumor.

6.4 What Is Needed to Restore the Economic Conditions of Regions Near the FDNPP

In sum, there are aspects where false rumors and poor reputation are causing damages to the sales of products produced in regions near the FDNPP. On the other hand, there are also conditions where other factors than reputation such as consumers' radiation knowledge and having children are affecting the consumer reaction to cause damages in the sales of these regions.

We found through our investigation that the reason for consumers to avoid buying food and beverage products from regions near the FDNPP is partly related to the behavior of consumers not seeking the level of risk involved with radioactive contamination of these products and believing whatever information that is available to them.

However, we also learned that some consumers have a negative reaction toward products of these regions because they cannot solely trust the current national food safety standards for radiation thinking that such safety standards are not omnipotent in all situations. It is likely that the avoiding behavior of such consumers is not only a result of bad rumors.

Therefore, to mitigate the effects of reputation damage and to restore the sales of products from regions near the FDNPP, the government needs to conduct policies that will enhance activities to obtain and spread scientific evidence to alleviate and remove fear associated with radioactive contamination. Furthermore, the government should create more opportunities for the consumers to learn and find out the risk involved with products from these regions. Although it is already conducted by the government, I believe the continuation of releasing information of the results of the radiation tests performed on the products of regions near the FDNPP and providing education programs on radiation and radioactive materials for the consumers is important for alleviating the effects of the ongoing reputation damage.

MIX
Papier aus verantwortungsvollen Quellen
Paper from responsible sources
FSC® C105338

If you have any concerns about our products,
you can contact us on
ProductSafety@springernature.com

In case Publisher is established outside the EU,
the EU authorized representative is:
Springer Nature Customer Service Center GmbH
Europaplatz 3, 69115 Heidelberg, Germany

Printed by Libri Plureos GmbH
in Hamburg, Germany